Commercial Floriculture

This book explores the industry of floriculture in its comprehensive nature. All commercially important plants cultivated have found place in the world, and have been classified on the basis of specific genera or families, which are also described. Informatory chapters on the floriculture industry, including soil and climate, manures and fertilizers, plant propagation, seed production, greenhouses and postharvest management have also been included.

With its erudite account of commercial floriculture, the book will be a handy reference for floricultural students, faculty and farmers.

Dr. S.K. Chattopadhyay a renowned authority on postharvest technology and corporate farming, has had a brilliant academic record. He had worked with various institutes and has participaated in many National and International Conferences. He had presented many papers of international repute in national as well as international seminars. As prolific writer, he authored a number of books on agricultural science and technology.

Commercial Floriculture

S K Chattopadhyay

2024

Gene-Tech Books

New Delhi - 110 002

ISBN 9788189729738

Published by : **GENE-TECH BOOKS**
4762-63/23, Ansari Road, Darya Ganj,
NEW DELHI - 110 002
Ph.: 011-43003222
e-mail: genetechbooks@yahoo.co.in

PRINTED IN INDIA

Preface

Floriculture has finally begun to form an important component of commercial agriculture; and with the increasing awareness of its potential, more and more people have become associated with this industry.

Floriculture today does not only cater to the needs of the local cash-crop industry, it has also begin to form a significant part of the economy in terms of the revenues it generates. Not only are its products used in ornamentation, but are also used in other industries like perfumery, and currently, it is in the stage of exploring possibilities of expanding its horizons to other vistas as well.

This book explores the industry of floriculture in its comprehensive nature. All commercially important plants cultivated have found place in the world, and have been classified on the basis of specific general or families, which are also described. Informatory chapters on the floriculture industry, including soil and climate, manures and fertilisers, plant propagation, seed production, greenhouses and post-harvest management have also been included.

With its erudite account of commercial florticulture, the book will be a handy reference for floricultural students, faculty and farmers.

Editor

Contents

1

Global Trends in Floriculture

Floriculture is the art and knowledge of growing flowers to perfection. It has become a remunerative enterprise all over the world. It needs intensive care and is suitable for small and marginal farmers especially around the cities. There is great potential for export and distant markets. Floricultural crops like roses, gerberas, carnetions etc. are grown in green houses. The open field crops are chrysanthemum, roses, gaillardia, lily, mary gold, aster, tube rose etc. Also there is great scope for cultivation of cut greens. Floriculture industry promotes activities such as tissue culture, grading and packing units, cold storage etc.

The flower industry comprises cultivation and trade of cut flowers, cut foliage, potted plants and bedding plants. The main representatives of cut flowers are: the rose, chrysanthemum, gladiolus, carnation and lilies. Potted plants and cut flowers have an almost 80 per cent share of the world trade in ornamental plant products. In the flower industry, significant changes are occurring in competitive relationships worldwide. A number of traditional markets are displaying signs of saturation and at the same time new markets are developing in some parts of the world. It is expected that per capita consumption and production will go up worldwide.

There is a dilemma in obtaining clear insight into the prospects for the floriculture industry. In almost every country in the world there is some florist industry but in many countries the available figures are not accurate.

The cultivation of commercial flowers is a lucrative business. The world trade in cut flowers is worth about 14.1 billion annually and total import of cut flowers into Europe alone is about 1.04 billion. The exporting countries are Holland, France, Italy, Israel, South Africa, Columbia, Thailand and Kenya. The contributions of India in this field is negligible.

The Asian flower industry has the potential to become the leading flower industry in commercial floriculture worldwide. Looking to the future, the Asian flower industry could rival, if not surpass, the size and scope of the European flower industry which presently dominates global commercial floriculture.

Flower Trade

The flower industry comprises the cultivation of and trade in cut flowers, cut foliage, potted plants and bedding plants. The main representatives of cut flowers are: the rose, chrysanthemum, carnation and lily. Potted plants and cut flowers have an almost 80% share of the world trade in ornamental plant products.

In the flower industry, significant changes are occurring in the competitive relationships worldwide. European integration, democratisation in Eastern Europe and the liberalisation of world trade in the context of the GATT negotiations will have consequences for international trade. A number of traditional markets are displaying signs of saturation while, on the other hand, new countries are trying to get their place on the market. At the same time, we see new developing markets and in some parts of the world it is expected that per capita consumption will go up. These many changes make it difficult to obtain clear insight into the prospects for the floriculture industry. Because of the sometimes highly changeable exchange rates between currencies, especially the dollar, the currebetween currencies, especially the dollar, the currencies of the original data sources are used.

Developments in Demand

Cultivation of cut flowers and potted plants is widely spread throughout the world. It is included in the statistics of 145 countries. In addition, cut flowers are extensively grown in many countries on small outdoor plots. Cut flower acreage and production value in the wormall outdoor plots. Cut flower acreage and production value in the world are increasing. Based on the seventeen most important production countries, it is currently estimated at about 60,000 hectares.

Demand for cut flowers

World consumption of cut flowers and plants can only be given in approximate figures. World cut-flower markets are growing at a current rate of 6-9% per year. The total consumption in 1985 was about 12.5 billion dollars. In 1990, the consumption rose to about 25 billion dollars. Due the change in the exchange rate between the dollar and the guilder, this growth is spectacular. In the nineties, the growth continues. In 1995, the total world market was about 31 billion dollars. Taking developments in production, imports and economic variables into account, consumption of cut flowers is to be expected to rise to 35 billion dollars.

In international terms, the consumption of cut flowers is concentrated in three regions: Western Europe, North America and Japan. The highest growth is expected in Japan and the USA. The West European market is becoming saturated. As a result, its cut flower consumption share is declining. New markets are emerging in the Eastern European countries. In one segment of the market, cut flowers are becoming part of the consumption patterns of people with high incomes. In Japan, domestic consumption of cut flowers will become more regular then in the past. In the past, consumption egular then in the past. In the past, consumption was mainly based on special occasions and institutional consumption. This pattern is changing to a more Western style. Consumption is rising not only in Japan.

In other Asian countries with rising expendable incomes and a flower-minded culture, consumption will go up. If economic development keeps accelerating in Latin America, a strong demand increase is also to be expected here. Growth in cut-flower consumption is greatly dependent on the economic development of the different parts of world and of course on flower-minded culture.

Demand for potted plants

Worldwide consumption of potted plants is increasing rapidly. The total consumption in 1990 was about 14.2 billion dollars, some 21% higher than in 1985. In 1995, this market already increased to about 19 billion dollars. Further growth is expected due to the growing expendable income in a lot of regions in the world. The US, with about one third of the consumption, has the largest share of the total world consumption, followed by Germany (about 20%), Italy and France.

Export/import of ornamental plants

The value of world import/export is increasing every year. In 1982, the total value was about 2.5 billion dollars. In 1996, the total value was alrea 2.5 billion dollars. In 1996, the total value was already 7.5 billion dollars. Although the growth is in nominal value, the figures show a rapid increase in the flows of horticultural products. The biggest import markets are Germany, USA, France and the United Kingdom. The world's largest exporter of horticultural products is by far the Netherlands, which is responsible for about 60% of international export.

The international trade in cut flowers and potted plants is growing every year. From the total production of cut flowers and potted plants, only a small part is exported to the international market. About 75% of the international trade comes from within Europe. Germany alone accounts for 30% of the world imports of cut flowers. Besides Germany, the United States, France and the United Kingdom are big importers. The Netherlands is also a big importer but it re-exports most of its imports, primarily to Europe.

The total value of the world export of cut flowers rose from 1.25 billion dollars in 1982 to 3.6 billion dollars in 1995. The Netherlands, Columbia, Israel, Kenya and Ecuador are the biggest exporting countries in the world. The position of the Netherlands in very dominant in the total export. When you consider the export from the Netherlands to other European countries (1.8 billion dollars from the total export value of 2.1 billion dollars) as inter export value of 2.1 billion dollars) as interregional trade, the conclusion is that the international trade is very limited in relation to total world wide consumption. It is to be expected that the north-south axis will be important to the export market. Africa will increasingly export to Europe, and South America to the USA and Canada. Within Asia, there will be a growing interregional trade with upcoming countries like Malaysia, Thailand and the Philippines. Australia and New Zealand have possibilities on the niche market in Asia with high-quality products.

The total export of potted plants is smaller, amounting to about 1.75 billion dollars in 1995. Because of the high transport costs per unit, the export of this type of product is usually to neighbouring countries. The main exporters are the Netherlands, Denmark and Belgium.

Production

Production, consumption and international trade of cut flowers and potted plants in the following main cultivation regions will be described: Western Europe; North and South America; Africa; and Asia.

Western Europe

There are no reliable figures about the domestic production values of cut flowers and potted plants in most EU countries. Figures for both area and production value are not very accurate. In many cases, the different statistics give different figures. Nevertheless, it is possible to give an indication of the main production countries in the EU by looking at the area under production and the latest production figures from the EU commission.

Table 1 shows the production area and trend in production area of the most important EU countries. It shows that The Netherlands has the largest area under production: more then 8,000 ha in 1996. This country also has the largest area under glass, with very intensive production. Italy had almost the same area in 1994, although the trend shows a decline in area. Although the total area in Italy is declining, there is a growth in production area under glass/plastic.

Table 1. Area under production of flower and foliage crops in the most important production countries in the EU

Country	Area	o.w. Glass/plast	Year
Netherlands	8004	5556	1996
Italy	7654	4402	1994
Germany	7066	2755	1996
United Kingdom	7127	999	1996
Spain	4325	2369	1994
France	3795	1747	1990
Belgium	1642	542	1993
Greece	990	-	1995
Denmark	683	330	1994

Other leading countries are Germany, the United Kingdom, Spain and France. Excepting Italy and France, we see an increase in the number of hectares under production. The growth of the sheltered area was the biggest in Spain. From 1980 to 1990, this area grew from 1,270 ha to 3,068 (242%). After 1990, there is decrease in area to 2,369 ha in 1994.

A large total area of flower and foliage crops does not automatically imply great production value. Differences in yield per hectare are considerable. For example, the yield per hectansiderable. For example, the yield per hectare in the Netherlands is 70% higher than in Italy. One explanation for this is that a greater share of production is carried out under glass. But there are great differences in yield even with production under glass, due to differences in production techniques. In France, Italy and the Netherlands, more than 60% of the total production is carried out under glass. In countries such as Germany, Denmark and the United Kingdom, this figure is under 30%.

Table 2. Production values of flower and ornamental plants of the EU-15 in 1980-1995

Country	Production value 1980	Production value 1995	Index (1980=100)
The Netherlands	1070	3429	320
Italy	681	1771	260
Germany	877	1401	160
France	745	937	126
Spain	154	564	366
United Kingdom	174	419	241
Denmark	156	346	221
Belgium	140	278	199
Sweden	122	158	130
Greece	30	160	533
Austria	48	108	225
Finland	70	104	149
Portugal	-	80*)	*)
Ireland	-	30*)	*)
Luxembourg	-	20*)	*)
Total	4028 1)	9678 2)	240

*Notes: 1) EU-12. 2) EU-15. *) Including tree nursery production. **) 1 ECU = 1.3 dollar, 1995. -) Not available.*

In 1995, the total production value was almost 10 billion ECU, which is about 13 billion dollars. In 15 years, production value went up by 240%. Besides the influence of inflation, there is an increase in production due to more intense cultivation per hectare.

North and South America

The North American growers supply about 90% of the domestic market. The quality of the extensive production, the low dollar exchange rate and the strict phytosanitary requirements for import have put the American growers in a favourable position. Floriculture and environmental horticulture is the fastest growing segment of US agriculture. The total production value is about 11 billion dollars in 1996. Most fresh cut flowers in the USA are produced year round and protected by a structure of some sort, such as around and protected by a structure of some sort, such as a greenhouse in which the environment is carefully controlled or an overhead structure providing shade and protection from the wind.

Table 3. Area under production of flower and foliage crops in the most important production countries in the USA and Latin America

Country	Area	o.w.Glass/plast	Year	trend (1985=100)
USA	15522	4532	1995	157
Mexico	10000		1994	
Columbia	4200		1995	
Costa Rica	3600		1994	
Ecuador	1620		1996	
Dominic.Rep.	400		1995	
Peru	200		1994	

Exact figures about the production of ornamental products in Latin America are not available. Its is clear, however, that the production volume is growing very quickly in this region. Due to the good climate conditions, the input of capital and knowledge, this region has reached a good export position to both the United States and the European Union. The two most important producers and also exporters are Columbia and Ecuador. Traditionally, Columbia was the biggest producer in this area, but Ecuador, although a relative newcomer, has a lot of potential to take over this position.

Columbia has about 2,500 ha of carnations and 1,500 ha of roses. This country's excellent position is due especially to the good climatic conditions, constant length of day and stable temperature, along with cheap land and labour and a favourable geographic location with respect to the North American market. However, due to rising production costs, low quality, diseases in certain production areas and a bad eases in certain production areas and a bad environmental image, this position is under pressure. In Ecuador, the area under production is growing explosively, increasing by hundreds of hectares per year. Strong production factors here include the climate, the considerable amount of sunlight, sufficient good-quality water and cheap land. In recent years, foreign capital from both Colombia and Europe has been invested in Ecuador. Production is especially focused on roses with long, thick stems. This product finds it way to the Russian market, where people are willing to pay good prices for it.

Africa

Cut-flower production in Africa is highly competitive to any professional grower in the world. Especially with regard to the European growers the competition is becoming fierce. The first cut-flower nurseries were established in Kenya in 1969, which originally is a tea and coffee producing country. Nowadays Kenya is the largest African cut-flower grower, followed by Zimbabwe, Morocco and South Africa. Other promising countries on the market are Zambia, Malawi, Tanzania and Uganda.

Traditionally, family farms growing cut flowers are a minority in Africa. The professional companies consist of large-scale nurseries owned by foreign investors, banks, wealthy individuals, cut-flower growers from Western Ewealthy individuals, cut-flower growers from Western Europe and local governments. At the major nurseries, the number of employees varies from a hundred to even thousands of people. The managers of these farms are often recruited from England, the Netherlands, Germany or Israel. On the other hand, small farms have also been established by local people in recent years. These countries benefit from a developing infrastructure, knowledge, transport facilities and entry to especially the European market due to trade agreements with the EU.

Bulk flower production is intended primarily for export to the European market, although there is increasing export to the Asian market as well. The quality of the flowers is very good at the time of harvesting but the products lose quality due to poor transport conditions. Africa used to produce primarily carnations, roses and summer flowers. Nowadays the growing of roses is increasing, because higher prices can be obtained on the export markets. The assortment is focused on the European market because high transportation costs oblige growers to strive for a high value per kilogram.

Due to the political environment in certain countries, ongoing investment in floriculture is under pressure. The future of floriculture in Africa greatly depends on the political circumstances in the near future.

Table 4. Area of flowers (and roses) in Africa (ha)

Country	Roses		Total
	1996	1997	
Kenya	425	585	1280 (1995)
Zimbabwe	300	360	940 (1995)
Ivory coast			690 (1995)
Morocco			427 (1992)
Tanzania	60	80	
Uganda	53	80	
South Africa	60	70	
Zambia	40	65	
Malawi	16	18	
Ethiopia	0	4	
Total	954	1262	

In one year alone, the total area under roses grew by some 30%. Especially in Kenya, the growth in the area of roses is very high. Four to six large-scale companies represent about 60% to 70% of the total production. A lot of the total area is in the open field or under sheltered conditions. But cultivation under plastic will increase due to better product qualities. With the establishment of more advanced greenhouses, there are increasing possibilities for growing a wider assortment of flowers. In addition, growers are more and more focused on growing all year round.

Growers in South Africa are focused on production for the local market. Other production countries grow flowers mainly to meet Western European consumer demands. The African countries make use of certain market opportunities, such as supplementing the European supply during the winter and cultivating species which are hard to g and cultivating species which are hard to grow in Europe or which require too much energy due to their long growing times. Although the African growers have an advantage in terms of production costs, they are worried that the fear of pesticides and the demand for environmentally friendly flowers will put pressure on production.

Since, as mentioned before, the majority of African flower production is exported to Europe, the volume and value of export gives an indication of production. Originally, export from North Africa was focused on roses. Due to fierce competition from other areas, the export focus has shifted to carnations. In 1991, about 60% of the export value was due to the export of roses. In 1996, this fell back to about 30%, the same share as carnations, which had a share of 5% in 1991. Besides roses and carnations, chrysanthemums are an important export product.

Export from West Africa is very limited at the moment. More significant is the growth in exports from Central, East and Southern Africa. About 90% of the total export comes from this region. In 1986, less then 3% of the volume came from roses. In 1996, the share in volume has risename from roses. In 1996, the share in volume has risen to almost 50%. The growth in export mainly comes from roses and summer flowers. Most of the flowers from Africa find their way to The Netherlands, England and France. In Europe, export goes directly through wholesalers. Moreover, flowers are sold through the Dutch auction system as well. More and more we see the tendency to bypass the Dutch auctions in favour of direct selling to retail organisations and wholesale markets. It is to be expected that separating the product flow and the commercial flow will change the traditional role of the auction in the near future.

Asia

The main production centres are:

- Japan, growing for the domestic market;
- Israel;
- India and;
- Southeast Asian countries such as Thailand and Malaysia, and Korea and China.

Along with the Netherlands, the US and Italy, Japan is a major producer of both flowers and potted plants, but its production is mainly for the domestic market. In 1994, the value of production was about the same as the Netherlands and the USA. The production of both potted plants and cut flowers is growing rapidly. Both have doubled in space in ten years to about 18,000 ha of flowers and about 2,000 hectares of potted plants. The limited but rapidly growing international trade provides a strong stimulus to domestic cut flower consumption. Domestic consumption is also benefi cut flower consumption. Domestic consumption is also benefiting from this.

Only one third of the cut flowers are grown under glass or plastic. The importance of protected cultivation is increasing. A shift is taken place form seasonal to year-round production. The main cut flower is the chrysanthemum, followed by the carnation and the rose.

In Israel, the total area increased in the early nineties. During the past four years, however, the total area has remained stable at about 1,900 ha. About 70% of the cultivation is done under glass, plastic or other protective cover. Israel is well known for its product innovation and knowledge transfer. It is the fourth largest exporter, mainly to the EU. The reduction of import duties by the EU gives Israel extra export possibilities. Due to the economic situation and the high exchange rate of the dollar, its export position is under pressure.

As a result of political changes in India, many entrepreneurs are changing to floricultural activities, which have become a high priority for the Indian government. Certain areas of India have great potential for floriculture:

— sufficient winter and summer sunshine;

— high temperature;

— good soil quality;

— good water quality;

— different climatic zones for different type of products;

— low labour and investment costs.

The sector is not yet very well organised with respect to the level of cultivation knowledge and post-harvest activities. The quality level is not sufficient to meet the high standards of the European and Asian markets, although progress is being made and the total export volume is rising. Export is hindered by high freight and tariff costs for shipment to Europe and insufficient cooling facilities at the airports.

Throughout Asia, the production of ornamental products is rising. In traditional production areas such as Thailand and Malaysia, production is more and more orientated to the Japanese and European marketed. In Thailand, more than 7,000 ha is under cultivation for cut flower production. Production is increasing in the newly developing countries such as Vietnam and China. In China, almost 60,000 ha are used for flower and foliage crops. In a period of three years, production increased from about 100 million stems to about 400 million stems. The development of the sector is just in an early stage. For example, standard quality, post-harvest technology and management skills are not at the required level at the moment. ls are not at the required level at the moment. Most of the production is for domestic consumption, but there is increasing export to mainly ASEAN countries for special products like Bonsai and potted plants.

Consumption of Floricultural Products

One can distinguish three main consumption markets for floricultural products: Western Europe; North America; and Japan.

Western Europe

Western Europe accounts for about half of the world's cut flower production. In terms of purpose, European consumption of floricultural products is very heterogeneous.

Consumption patterns differs from country to country in Europe. Flowers are generally purchased for gifts, special occasions such as birthdays and weddings, and home decoration. Consumers expect flowers to be colourful and beautiful. The emotional aspect is important. The European market is a high-quality market. Consumers want to purchase fresh products with long vase lives. The scent of flowers will become increasingly important.

Table 5. Consumption of cut flowers 1990-2000 in some European countries (ECU million)

	1990	1995	2000
Germany	2547	3210	3236
Italy	2156	1638	1736
France	1366	1535	1884
United Kingdom	928	934	1093
The Netherlands	492	595	631
Spain	386	441	569
Belgium/Luxembourg	270	404	497
Austria		310	362
Sweden	268	310	362
Finland		248	289
Norway	220	241	294
Denmark		199	243
Greece	145	138	206
Portugal		101	158
Ireland		44	52
Switzerland		611	680

The consumption of cut flowers in the EU has grown rapidly in recent years. Total consumption in 1995 is estimated at 11 billion ECU. The differences between countries are big. Germany dominates flower sales, followed by Italy and France. In recent years, the consumption of flowers has grown in all the countries of Europe except Italy.

In 1995 the total flower consumption in the aforementioned European countries was already about 11 billion ECU. This market will grow to almost 12.5 billion ECU by 2000, an increase of 12%. When consumption in Eastern Europe, which is gradually growing, is taken into account the European market will be the biggest consumer market in the world.

The non-EU country Switzerland has by far the highest per capita expenditure. The per capita differences are big. In Portugal, the per capita expenditure is about 10 ECU while in Finland the average is about 50 ECU.

Table 6. Consumption per capita of cut flowers 1990-2000 (ECU)

	1990	1995	2000
Germany	32	40	41
Italy	36	29	30
France	25	27	34
United Kingdom	16	16	19
The Netherlands	33	39	42
Spain	10	11	15
Belgium/Luxembourg	25	40	50
Austria		46	53
Sweden	32	35	42
Finland		50	58
Norway	52	54	69
Denmark	31	38	47
Greece	15	13	20
Portugal		10	15
Ireland		11	10
Switzerland	7070	87	102

North America

The definition of floriculture differs from country to country, making it difficult to compare statistics. In the USA, the floriculture sector includes crops such as cut flowers, cultivated greens, potted flowering plants, potted foliage plants and bedding and gardening plants. In the USA, cut flower consumption is not as dominant as in Europe. Over the past five years, the total wholesale value increased about 15%, representing an average growth of about 3% per year. The quantity of domestic cut-flower production decreased in the same period. Other international suppliers filled up the gap. The most important flowers are carnations, chrysanthemums, gladioli and roses. For the most important flowers, the average import share is between 70%-90%. Besides the increased consumption of cut flowers, the consumption of potted flowering and bedding plants also went up.

Fresh cut-flower sales to consumers traditionally have been made by retail florists for use in connection with weddings, funerals, gifts, special occasions, etc. In the last five years, there has been a large increase in sales of flower and potted plants through mass marketers such as super markets, garden stores, mass marketers such as super markets, garden stores, flower stands and street vendors. They offer fresh domestic and/or foreign products, depending on price, availability and quality.

Japan

Japan is one of the largest consumers of cut flowers. Both the total consumption of flowers and the average peth the total consumption of flowers and the average per capita consumption are high. Individual flower consumption is lower than in the other major industrialised countries. The reason for this is that 70% of the total consumption derives from business concerns that handle ceremonial occasions such as weddings and funerals, from corporate gifts, from floral decorations in hotels, and so on. Furthermore, the comparatively high costs of a single flower in Japan tend to disincline consumers to buy. The custom of buying flowers for one's own use is relatively unknown in Japan. In recent years, there is a trend to a more European life style. Consequently, the percentage of the total consumption represented by individual purchases is likely to continue rising. With a total consumption of 5 billion dollars, this major market is attractive to suppliers all over the world.

Japan's flower market has changed drastically in the last ten years. Although institutional demand has fallen due to the declining economy, non-commercial demand is growing. Currently, institutional demand has fallen back from about 40% to 30%. Non-commercial demand, both personal and regional, is becoming an established part of the market. Personal consumption is about $ 130 per household and about $ 100 for gardening products. ehold and about $ 100 for gardening products. According to the Japanese Flower Promotion Centre, demand for cut flowers will rise by around 60% by 2005 due to the growth in the casual flower and personal gift markets. Consumers will look for a wider variety of products as personal demand grows. The per capita consumption will rise from 50 stems in 1993 to 76 stems in 2005. This means a average growth of about 4% per year. Most of the growth will come from domestic production, although import will increase to meet the growing demand.

Trade of Ornamental Products

Latin America is the biggest export region for ornamental products in the world. Total export is about SFR 950 million, which was about 750-800 million dollars in 1996. About 60% of the total export of this region goes to North America. More than 90% of this export to North America is cut flowers. Export countries include Colombia and Ecuador, primarily for cut flowers, and Costa Rica and Guatemala for potted plants.

The second biggest export region is Europe, with a total export of approx. SFR 720 million. Although the Netherlands is considered to be the biggest exporter of flowers and considered to be the biggest exporter of flowers and plants, one can see that its role at world regional level is not so impressive. This is due to the fact that most of its export is oriented to the countries within the EU.

Most of the export of the EU still comes from the Netherlands, but this position is mainly based on the export of bulbs, in which it have a very strong and unique position. Within Europe, the Netherlands exports about SFR 4.1 billion worth of products, which represents 80% of the total trade of ornamental products in the EU. The products are mainly flowers and potted plants within Europe and bulbs interregionally.

Africa, in particular, has very rapidly established a position on the export market. In a period of ten years, Africa managed to increase its export, especially to Europe, by more than 300%. Total export is SFR 340 million to Europe. Africa's export position with respect to other world regions is rather insignificant. Leading export countries are Zimbabwe and Kenya, which account for more than 80% of the export. In the fourth place comes the Middle East, which actually means Israel. Total export was about SFR 257 million in 1996, of which 80% was cut flowers (mainly roses).

Asia plays a rather small role in the international trade. Most Asian trade is within Asia itself. Only for special products like orchids (Thailand) and Bonsai products (China/Japan) is thereThailand) and Bonsai products (China/Japan) is there is small export position, amounting to about SFR 120 million, of which 80% goes to Europe.

Import from Regions

Europe is the biggest import market for ornamental floricultural products. Total import is about SFR 1.2 billion. North America is the second largest import market for ornamental floricultural products. Outside North America, the USA imports about SFR 1 billion worth, which is about 760 million dollars. The position is just the opposite when one considers only the trade of cut flowers. For cut flowers, the North American market is the biggest import market.

International Competitiveness

In this section, the specific aspects of cut flower and pot plant production centres are examined within Porter's framework. Porter's analysis is based on the theory of the industrial ecoframework. Porter's analysis is based on the theory of the industrial economy. Porter came to the conclusion that the competitive strength of nations is determined by four interrelated factors which shape the national environment. Because he diagrams these factors as the four points of a diamond, he calls this structure the diamond of a nation. Nations are most likely to succeed in industries where the national diamond is the most favourable. The four determinants of Porter's diamond are:

1. *Factor conditions.* The nation's position in terms of production factors such as skilled labour and infrastructure which are necessary to compete in a given industry;
2. *Demand conditions.* The nature of domestic demand for the industry's product or service;
3. *Related and supporting industries.* The presence or absence of internationally competitive suppliers and related industries in the nation;

4. *Firm strategy, structure and rivalry.* The conditions in the nation governing the creation, organisation and management of companies and the nature of domestic rivalry.

Apart from these four basic determinants, Porter distinguished two additional factors:

5. *Chance.* Events that have an influence on the competitive positions of the industry, such as important technological breakthroughs and discontinuities in input costs caused by oil price shocks, wars, changes in exchange rates, etc. Consider what is happening now exchange rates, etc. Consider what is happening now in Asia and its impact on changing international product flows. A change in the export positions of these countries is to be expected;

6. *The role of the government.* The role of the government has an influence on the four determinants of the diamond and on the competitiveness of a sector.

Countries or regions can derive their competitiveness from a low cost price, a high level of investment, rapid innovations or their competitive positions and innovations in the past. Those which derive their strength from rapid innovations are the most effective competitors. These countries are not dependent on low cost prices or investments in innovations by other countries, but are strong enough to make innovations themselves. These innovations are the result of a good balance between the production factors, the economic variables, domestic demand and the networks. An active search is made for new opportunities and the best manner of confronting threatening developments.

In the cut-flower sector, the Netherlands is the only country which derives its competitiveness from rapid innovations. Although the great innovative ability in the Netherlands has been primarily concentrated on the suppin the Netherlands has been primarily concentrated on the supply side, a change to more awareness of the demand side is clearly noticeable. Stronger vertical integration to meet the demands of consumers and retailers, more effective logistics, higher quality products and environmentally sound production facilities are all current topics. Disadvantages for the Dutch sector such as expensive raw materials and labour and a less favourable climate are compensated for by strong points such as high productivity, a tightly-knit and efficient distribution network and a wide, innovative range of products.

The country's favourable position with respect to Europe, its capacity to invest, its good infrastructure and the comprehensive knowledge and network available to the sector constitute a good seedbed for innovations. The network in which the sector is incorporated is not only vertical but is also increasingly developing horizontally. Although the Dutch sector also competes internationally in the area of costs, it does not depend on low costs for labour, land, raw materials and capital. Its strength is based more on its ability to innovate rapidly. This is reflected in its productivity, quality, range of innovations, high professional skill and state-of-the-art technology.

In markets where cost prices are important, the Netherlands will experience increasing competition from African countries such as Kenya, Zambia and South Africa, and from Latin Americania and South Africa, and from Latin American countries like Columbia and Ecuador. By continuing to innovate in terms of both technology and range, the Netherlands can maintain a leading position for cut flowers and potted plants in Europe.

African countries are strong in terms of basic factors. the climate is favourable and there is and adequate supply of cheap land, water and labour. On the other hand, there are no innovative stimuli from knowledge infrastructures, strong networks and domestic markets. Also, unlike the Netherlands, countries in Africa do not have the capital to develop knowledge and technology on a large scale themselves. They are greatly dependent on knowledge provided by suppliers, retailers, and other third parties. As a result, they are only able to produce products which require limited technology and knowledge. They can offer a reasonable volume of these products at a low cost price on the world market.

A further growth in competitiveness is possible if capital becomes available and they are willing to invest. This development is hampered, however, by political instability, which has its effects on the innovation process. Moreover, growth in production depends on the availability of cheap air freight facilities. Vertical integration is also seen in Africa. Because of good coordination on the suppliers' side, Africa remains strong in products with high volume and low cost price.

The countries in Latin America derive their strength from a high level of investment. In many cases, similar to the Netherlands, a great deal is invested in technology. With the input of a lot of foreign capital, the production area is increasing rapidly. High-quality factors are being strengthened with the aid of investments. Thanks to strong basic factors and advantages of scale, standard products can be readily supplied through the international market channels. Latin America has the potential for rapid growth in the coming years. Sufficient money is available for investments in knowledge, capital goods and range of products. As a result of increasing production and additional air transport facilities, sales to Europe will continue to climb. The cost price level has great influence on prospects. The high expense of transport must be offset by a low cost price. Because of the high dollar exchange rate, export to North America will expand more easily. The European market will only be reachable if the cost price is kept down and sufficient affordable air freight facilities are available.

Distribution Changes

Due to changes in worldwide economic circumstances, the international trade in ornamental products will be changed in the coming decade. What are the driving forces behind these changes? In his book, John Kotter demonstrates in simple form what one can regard as the driving forces behind the changes currently taking place in industry, changes which are also affecting horticultural agribusiness.

First of all, the rapid technological changes. It looks as if we are going through an information technology revolution. Communications are getting faster, better and cheaper, and the same goes for transport. A worldwide network links the world day and night, with high expectations regarding the role of this network in many different areas. Almost everyone uses a computer. Millions of people sit at the screen again in the evening to look for information connected with their work or their hobbies. Because more people are working and there are increasing opportunities to be active from home, consumers will want to spend less time on domestic work, which has consequences for the manner in which products are offered on the market. It is expected that information technology will play a major role here. There are high expectations regarding the turnover that can be achieved through the Internet.

Secondly, international economic integration. The implementation of the GATT rounds, the establishment of large free trade zones, interlinked currencies and the steadily expanding circuit of worldwide flows of capital ensure a more balanced economic development. Some economists say that we are back at the start of a Kondratieff cycle in which we will experience a long period of economic growth. We increasingly have to deal with customers who operate on a globally have to deal with customers who operate on a global scale. Our competitors will increasingly operate on a global scale too. The assets of large institutional investors are invested all over the world. In some parts of the world large investors are putting capital in floriculture.

Third, saturated markets. Growth in the traditional developed countries is slower because of the increasing saturation of the markets. This forces companies to orientate themselves more internationally. At the same time, a farreaching deregulation of markets is taking place, as a result of which international flows of goods can proceed more easily.

Lastly, free market mechanisms. After the collapse of the Berlin Wall, Eastern European countries have experienced an accelerated orientation towards freer market mechanisms. The same applies to Asia, where a country like China is gradually making steps towards the introduction of a market mechanism. At the same time, the government is in retreat in countries where there was already a form of market economy, so that more market mechanisms are being introduced in production and services.

The items listed above contribute to a further globalisation of markets. The entry of many enterprises to this market therefore leads simultaneously to more competition. This means more opportunities, but also extra risks. Having customers for opportunities, but also extra risks. Having customers for your product is no longer only dependent on the lowest cost price. Before you know it, a supplier with a lower price will be there. Retailing on markets will become more complicated. Service, quality and reliability are essential factors for securing your position on such international markets.

2

Cut Flower Production

Floriculture is concerned with the cultivation of flowering and ornamental plants for gardens and for floristry. The development plant breeding of new varieties is a major occupation of floriculturists. Floriculture crops include bedding plants, flowering plants, foliage plants, cut cultivated greens, and cut flowers. As distinguished from nursery crops, floriculture crops are generally herbaceous. Bedding and garden plants consist of young flowering plants and vegetable plants. They are grown in cell packs, in pots, or in hanging baskets, usually inside a controlled environment, and sold largely for gardens and landscaping. Geraniums, impatiens, and petunias are the best-selling bedding plants.

Flowering plants are largely sold in pots for indoor use. The major flowering plants are poinsettias, orchids, florist chrysanthemums, and finished florist azaleas. Foliage plants are also sold in pots and hanging baskets for indoor and patio use, including larger specimens for office, hotel, and restaurant interiors. Cut flowers are usually sold in bunches or as bouquets with cut foliage. The production of cut flowers is specifically known as the cut flower industry. Farming flowers and foliage employs special aspects of floriculture, such as spacing, training and pruning plants for optimal flower harvest; and postharvest treatment such as chemical treatments, storage, preservation and packaging. The production and marketing of cut flowers can be a profitable undertaking for small-scale growers. Flowers can be grown by themselves, or can be rotated with vegetables.

Cut flower production offers several opportunities and advantages:

- It is a way to diversify your operation
- It can be an excellent income supplement
- Flowers can be relatively easy to grow and inexpensive to produce
- Flowers can add beauty to your farm and home
- There is a good market for cut flowers
- You can grow varieties which are not readily available in retail outlets
- Flowers can attract bees and other beneficial insects
- Growing flowers can be therapeutic for growers with health problems.

Cut Flower Production in India

In India, with changing life styles and increased urban affluence, floriculture has assumed a definite commercial status in recent times and during the past three decades particularly. Appreciation of the potential of commercial floriculture has resulted in the blossoming of this field into a viable agri-business option. Availability of natural resources like diverse agro-climatic conditions permit production of a wide range of temperate and tropical flowers, almost all through the year in some part of the country or other. Improved communication facilities have increased their availability in every part of the country. The commercial activity of production and marketing of floriculture products is also a source of gainful and quality employment to scores of people.

Inspite of the long and close association with floriculture, the records of commercial activity in the field are very few. The information on the area under floriculture and the production generated is highly inadequate. As commercial floriculture is an activity which has assumed importance only in recent times, there are not many large farms engaged in organised floriculture. In most part of the country flower growing is carried out on small holdings, mainly as a part of the regular agriculture systems.

Production Areas

The estimated area under flower growing in the country is about 65,000 hectares. The major flower growing states are Karnataka, Tamil Nadu and Andhra Pradesh in the South, West Bengal in the East, Maharashtra in the West and Rajasthan, Delhi and Haryana in the North. It must, however, be mentioned that it is extremely difficult to compute the statistics of area in view of the very small sizes of holdings, which very often go unreported. This perhaps would be the reason for unrealistically small areas reported for floriculturally active states like Maharashtra, Uttar Pradesh and Madhya Pradesh. More than two thirds of this large area is devoted for production of traditional flowers, which are marketed loose e.g. marigold, jasmine, chrysanthemum, aster, crossandra, tuberose etc. The area under cut flower crops used for bouquets, arrangements etc. has grown in recent years, with growing affluence and people's interest in using flowers as gifts. The major flowers in this category are rose, gladiolus, tuberose, carnation, orchids and more recently liliums, gerbera, chrysanthemum, gypsophila etc.

Table 1. Area of flower production in India

State	Area (ha.)
Karnataka	19,161
Tamil Nadu	14,194
West Bengal	12,285
Andhra Pradesh	5,933
Maharashtra	3,356
Rajasthan	1,985
Delhi	1,878
Haryana	1,540
Madhya Pradesh	1,270
Uttar Pradesh	1,000
Others	2,166
Total	64,768

The production of flowers is estimated to be nearly 300,000 metric tonnes of loose flowers and over 500 million cut flowers with stem. In the case of production also, the estimates could be at variance from the actual figures as some of the flowers like rose, chrysanthemum, and tuberose are used both as loose flowers and with stem.

Almost all of the area reported here is under open field cultivation of flowers. Protected cultivation of flowers has been taken up only in recent years for production of cut flowers for exports. The estimated area in production is about 200 hectares, which is likely to increase to over 500 hectares by the year 2000. Recognising the potential for low cost production for export, in view of cheap land, labour and other resources, several export oriented units are being set up in the country. These projects, located in clusters around Pune in the West, Bangalore and Hyderabad in the South, and Delhi in the North, are coming up in technical collaboration with expertise mainly from Holland and Israel. More than 90 percent of these units are for rose production, on an average size of 3-hectare farm, while some projects for orchid, anthurium, gladiolus and carnation are also being set up. Nearly one third of over 200 proposed projects, have already commenced production and export.

Major Cut Flower Crops

Rose is the principal cut flower grown all over the country, even though in terms of total area, it may not be so. The larger percentage of the area in many states is used for growing scented rose, usually local varieties akin to the Gruss en Tepelitz, the old favourite to be sold as loose flowers. These are used for offerings at places of worship, for the extraction of essential oils and also used in garlands. For cut flower use, the old rose varieties like Queen Elizabeth, Super Star, Montezuma, Papa Meilland, Christian Dior, Eiffel Tower, Kiss of Fire, Golden Giant, Garde Henkel, First Prize etc. are still popular. In recent times, with production for export gaining ground in the country, the latest varieties like First Red, Grand Gala, Konfitti, Ravel, Tineke, Sacha, Prophyta, Pareo, Noblesse. Virsilia, Vivaldi etc. are also being grown commercially.

Gladiolus is the next most important cut flower crop in the country. Earlier it was considered a crop for temperate regions and its growing was restricted to the hilly areas, particularly in the north eastern region, which still continues to supply the planting material to most parts of the country. However, with improved agronomic techniques and better management, the northern plains of Delhi, Haryana, Punjab, Uttar Pradesh, as well as Maharashtra and Karnataka have emerged as the major areas for production of gladiolus.

Tuberose, a very popular cut flower crop in India is grown mainly in the eastern part of the country i.e. West Bengal, and also in northern plains and parts of south. Both single and double flower varieties are equally popular. Tuberose flowers are also sold loose in some areas for preparing garlands and wreaths.

The other main cut flower item is orchid. Its production is restricted mainly in the north-eastern hill regions, besides parts of the southern states of Kerala and Karnataka. The main species grown are Dendrobiums, Vanda, Paphiopedilums, Oncidiums, Phalaenopsis and Cymbidiums.

Among the traditional crops grown for loose flowers, the largest area is under marigold, grown all over the country. In most parts of the country only local varieties are grown for generations. African marigolds occupy more area as compared to the small flowered French types. Jasmine flowers in view of its scent are also very popular as loose flowers and for use in garlands and *Veni*. The major areas under this crop are in Tamil Nadu, Karnataka in South and West Bengal in East.

The varieties are mainly improved clones of *Jasminum grandiflorum,* J. *auriculatum* and J. *sambac*. The chrysanthemum, particularly the white varieties are much in demand as loose flowers during the autumn period of October-December when other flowers like jasmine, tuberose are not available for use in garlands etc. Among other traditional flowers grown in large areas are crossandra in southern states of Tamil Nadu, Karnataka and Andhra Pradesh and aster in Maharashtra.

Research work on floriculture is being carried out at several research institutions under the Indian Council of Agricultural Research and Council of Scientific and Industrial Research, in the horticulture/floriculture departments of State Agricultural Universities and under the All India Coordinated Floriculture Improvement Project with a network of about twenty centres. The crops which have received larger attention include rose, gladiolus, chrysanthemum, orchid, jasmine, tuberose, aster, marigold etc. The thrust till recently had been on crop improvement, standardisation of agro-techniques including improved propagation methods, plant protection and post harvest management.

In view of the fact that most of the cut flower production is being done under open field conditions, the research efforts generally relate to open cultivation. In recent years, however, technologies for protected cultivation and tissue culture for mass propagation have also received attention. A large number of varieties suitable for cut flower use, as well as garden display have been developed. Production technology, particularly the agronomic requirements and control methods for important diseases and insect pests have also been developed. Contribution by the private sector in research activities in floriculture is negligible.

Crop Selection

There are annuals, perennials, grasses, woody shrubs, trees and vines all suitable for use as cut flowers or other cut plant material. Flowers can be grown specifically for fresh use, dried use, for parts other than flowers such as seed pods, or any combination of the preceding. Start with species or varieties proven to grow in your area - Virginia's climates are as varied as its geography. Introduce other species on a trial basis before committing to large-scale plantings. An interesting industry trend is toward smaller acreage for cut flower production: 0.5 to 2 acres. Overproduction can be a problem. Most local markets can't accommodate the volume produced on 10 acres of cut flowers, no matter how beautiful they are!

Annuals will produce in the year planted and most are easily started from seed. Annual seed is readily available and lower in cost compared to perennials and bulbs. Tender or half-hardy perennials can also be grown as annuals. The choices for perennials are endless. Some traditional perennial crops that bring a good price include peonies (both fresh and dried), lily of the valley, and calla lilies.

Garden-type roses with fragrance are a possibility for the specialty florist market-also the heirloom roses such as cabbage and bourbons could fetch top dollar during the wedding season. Woody species can be used to extend production time to include early spring flowers (either naturally or on forced branches such as pussy willow or flowering apricot), summer flowers, late season berries, or ornamental stems such as the corkscrew willow harvested in the fall and winter. Relatively low maintenance and long-term production are benefits of including "woodies" in your field-grown cuts programme. Do keep in mind that it may take a few years for some species to reach marketable harvest size.

Both wholesalers and florists want to be the first with unusual or improved varieties. If you have the space, consider a larger product mix. A wide varicty of cuts allows you to service a greater percentage of the needs of your customers. Consistently test new crops. The ability to provide the floral wholesaler or retailer with unusual material will make you invaluable. Most wholesalers or retailers will pay for samples of a new variety. In the same vein, be aware of changes in consumer style and colour preferences.

Crop Production Methods

Site Selection

The ideal site for cut flower production is in full sun with wind protection, an irrigation source, and easy access. Raised beds are virtually required to ensure proper drainage (unless the soil is very sandy). Beds can be any length. Keep in mind that beds should be narrow enough for you, your employees, and/or customers to reach comfortably to the middle of the bed. A bed height of 4 to 8 inches should suffice, and drainage can be enhanced by burying drain lines 12 inches below the beds. Break up the hardpan or clay layer at least 1.5 to 2 feet deep with a subsoiler. Proper preparation of the beds is essential to optimum growth and yield.

Amend beds with plenty of organic material. Well-composted horse or cow manure is excellent. Be sure it has aged to the point where weed seeds are no longer viable. Incorporate lime, superphosphate, and other nutrients as recommended by the results of your soil test. If any of the beds are to lay fallow for a length of time, such as over winter, cover crops are excellent for increasing the organic matter content of the soil and reducing erosion. Recommended fall and winter cover crops for Virginia include hairy vetch, rye, barley, and crimson clover; or better yet, a combination of two or more of these crops. Summer cover crops can include soybeans, cowpeas, buckwheat, and sesame. Mulch has its proponents and detractors. Some growers insist on it for weed suppression and moisture retention. Others prefer to till between rows to control weeds. Because of the large variety of species grown as cut flowers, no single herbicide can be recommended.

Planting Material

Plant material can be acquired from a number of sources. Annuals can be direct sown into the beds or transplanted as plugs. Some annual species should be seeded in succession to continually produce as long as the growing season allows. Perennials are best started from plugs, whether seed or vegetatively propagated. There are many commercial sources for large plugs and starter plant liners. If perennial species are planted out in the spring, be sure they have been vernalised if that particular species requires it in order to flower that season.

The requirement of planting material to cater to the large area under flower crops, is largely met from domestic production. Since efforts to set up large commercial farms generally suffered due to lack of quality planting material in sufficient quantities, this aspect has received greater attention in recent years in the breeding centres, which are producing sufficient quantity of planting material. Most of the nurseries propagating planting material are in the private sector. In the absence of any mechanism to register nurseries, it is very difficult to ascertain their exact number, but at a very conservative estimate there are more than 100,000 nurseries, spread out all over the country, producing seeds and other planting materials for flower growers.

The states with larger numbers of nurseries include Maharashtra, West Bengal, Karnataka and Tamil Nadu. Most of the nurseries are small, with little or no improved facilities like mist propagation unit, green houses/net houses etc. For meeting the demand of flower seeds, several large seed companies have production units in Punjab, Himachal Pradesh and Jammu & Kashmir in the North, Karnataka in the South and West Bengal in the East. A few of the leading multinational seed companies have tied up with local seed companies or producers for custom production of seeds of their varieties. In the case of bulbous plants, most of the planting material is produced in the north eastern hilly regions of West Bengal and Sikkim, though for some crops, it is also produced in hilly regions of northern India. The introduction of a revised seed policy by the government of India in 1989 has enabled unrestricted introduction of many new and superior varieties into the country, increasing the variety in the floral basket.

Tissue culture has, in recent years, been recognised as an important tool in agriculture development. With its diverse climatic zones and qualified manpower, India is well placed to exploit the benefit of tissue culture based applications to floriculture crops. Most popular application of tissue culture has been micropropagation using *in vitro* technique for mass multiplication of planting material.

Tissue culture plants of ornamentals have found ready acceptance by the commercial growers and their production increased significantly from 130 million plants in 1985-86 to 680 million in 1994-95. At present 30 commercial tissue culture units with annual capacities of 0.5 to 15 million plants each are in operation, resulting in total capacity of about 110 million plants. While most of it is exported, a small percentage of cut flower crops like carnation and gerbera are finding good market within the country.

Spacing and Support

Optimal spacing varies between species. What seems adequate spacing for a row of perennials the first year may result in overcrowding the following year. Dense spacing can lead to higher incidence of disease as air circulation is limited. Conversely, too much space is an invitation to weeds and reduces yield per square foot. Note that dense planting often encourages longer stems.

The spacing of annuals varies by species, ranging from 4 to 6 inch centers to 1 foot x 1 foot. Be sure to thin rows to proper final spacing if direct seeded. Depending on the species[1] particular vigor, recommended spacing for perennial species ranges between 1 foot x 1 foot and 2 foot x 2 foot or 1 foot between plants and 2 to 3 feet between rows. Woody plants should be placed on 3-foot centers with more aggressive or larger species on 5-foot centers. Maintain moist conditions until plants are well established. Division is beneficial (or imperative) for many perennial species after the second year of production in order to maintain productivity.

Tall or relatively top-heavy species will require stem support. Rig beds with adjustable wire or plastic netting that can be raised as plants mature. Be sure the netting and supports are in place before the plants get too tall. It's extremely difficult to "retrofit" support without damaging the plants.

Many species, such as Phlox paniculata, benefit from pinching to encourage branching and obtain the maximum number of stems per plant. Pinch as soon as the plants are well-established and elongating. Leaving some of the crop unpinched can result in earlier flowering and larger flowers with the trade-off of fewer stems. Consult crop-specific references for timing.

Fertiliser Requirements

Fertiliser requirements differ from crop to crop. Some annuals, such as the annual sunflower (Helianthus annuus), are heavy feeders and require periodic fertilisation throughout the growing season. Fertiliser delivery methods range from broadcast or side dress application of granular fertiliser to fertigation (application of nutrients through the irrigation system). Incorporation of a slow-release fertiliser during Spring (not Fall) tilling will give young plants a jump-start. Periodic on-site monitoring of soil and irrigation water pH and soluble salts will be a tremendous help in designing and adjusting an appropriate fertiliser programme. Limited space does not allow a full discussion on fertiliser sources and recommendations.

Pest Management

Pests for field grown cut flowers run the evolutionary gamut, from powdery mildew, aphids and Japanese beetles, to rabbits, deer, and unscrupulous passers by. Fungicides, pesticides and shotguns are all of use in the battle for maximum yield. IPM (integrated pest management) is highly recommended as a money-saving and environmentally acceptable pest and pathogen control method for field cut flower production. If you are using only organic means or biological controls to produce your crop, tell your customers.

Crop Harvest and Handling

In warm weather, it is imperative that field-grown cut flowers be harvested early in the morning. Harvest when plant water status is high and temperatures and transpiration are relatively low. Wait until dew or other moisture has evaporated, if possible. Wet flowers and foliage are more susceptible to postharvest pathogens. Do not harvest when light level and temperature are at a maximum. Shading the freshly harvested material also helps maintain lower temperatures.

Harvesting at the proper stage of development for each species is very important. Too early, and some species may not open; too late can result in drastically reduced vase life. Harvest is the most labor-intensive aspect of cut flower production. Communication with your harvest workers is essential! Be clear about what is acceptable and what isn't to insure uniformity of the product.

Field and handling sanitation is just as important as it is in the greenhouse business. Keep fertiliser injector systems, harvest knives or shears, postharvest handling buckets, surfaces and coolers clean and sanitised. Do your cutting, grading and bunching in one "fell swoop" to eliminate excess handling which can increase the cut's exposure to pathogens and water stress. There are numerous ways to preserve and make use of surplus cuts - air drying, oven drying, silica gel, glycerin, etc.

Postharvest Handling

Proper postharvest care of your cuts is essential for maintaining high quality and a long vase life. The plant's life processes continue even after the stem is cut; respiration, transpiration, growth and development still happen. The cut stems and flowers remain sensitive to damage and disease. Floral preservatives and other additives are a necessary part of the postharvest process.

Cool water can serve to promote cooling of the stems. Warm water is useful if the cuts are under extreme water stress. Monitor water pH - acidic water (pH 3.0 to 5.5) inhibits bacterial growth helping flowers persist longer. Preservatives are also formulated to be effective at lower pH. Mixing your own postharvest preservative concoctions is not recommended. There are many commercial sources for flower preservatives, conditioners, hydrators, and ethylene inhibitors.

Ethylene is another important consideration affecting postharvest longevity. Flowers cannot be stored with any kind of fruit or vegetable. The ethylene produced by the fruit or veggie will result in premature floral senescence. Good ventilation and removal of dead and dying flowers are essential to maintain a relatively ethylene-free environment.

Grading, Packing and Delivery

There is no mandatory grading system for specialty cut flowers. General rules of thumb apply, however, emphasising uniformity: no greater than 10% deviation in stem length, relative uniform stem diameter, flowers of uniform size and stage of development. Ten stems per bunch is the standard for most species, with some species sold in fives or as singles.

There are myriad packing options - buckets, boxes, flowers held wet or dry; find out which are appropriate for the species you are growing. The majority of specialty cut flower growers use the indispensable 5-gallon plastic bucket. Be aware that some species such as snapdragon and gladiolus exhibit a geotropic response: stems laid flat will bend upwards, away from the gravitational pull, resulting in curved stems.

Vehicular and personnel requirements necessary for timely deliveries are often overlooked in the planning of a cut flower business. Things can get complicated, reducing efficiency and profits. One alternative is to deliver directly from the field in the morning. This works adequately for immediate delivery to local markets. However, if you need to hold the flowers for any reason, such as accumulating certain cuts for a larger wholesale market, cold storage facilities will be necessary.

There are many options available at a wide range of costs - built-in-place coolers, prefabricated cold storage units, or even modified refrigerated transportation units such as refrigerated semi-tractor trailers or ice cream trucks. If using an independent shipper, be sure your carrier is educated to the need to maintain temperatures between 35° and 40° F during transit.

Cut Flower Production Development

The availability of natural resources like favourable and diverse climatic conditions permit production and availability of a large variety of flower crops round the year. Cheap labour leads to reduction in production costs, increasing access of the consumer to good quality flowers at affordable prices, besides increasing our competitiveness in the export markets. Being a new concept in the agri-business, it took some time for scientific commercial flower production to take roots, but with the appreciation of its potential as an economically viable diversification option, its growth is slowly stabilising. The government also has, during the last few years, recognised floriculture as an important segment for developmental initiatives.

Model Floriculture Centres being set up in 11 major production zones, to serve as focal units for development in the region, have a mandate of making available quality planting material, new/improved production technologies and also to provide training in production and post harvest management. There are also special government programmes for area expansion in floriculture with state assistance. The National Horticulture Board, a major developmental agency for horticulture, also makes available finances as soft loan for setting up integrated projects for production and marketing. As mentioned earlier, the government is investing in improving the infrastructure for marketing in the domestic sector.

Production of cut flowers for exports is also a thrust area for support. The Agricultural and Processed Food Products Export Development Authority (APEDA), the nodal organisation for promotion of agri-exports including flowers, has introduced several schemes for promoting floriculture exports from the country. These relate to development of infrastructure, packaging, market development, air freight subsidy etc. The 100% Export Oriented Units are also given benefits like duty free imports of capital goods. All these efforts indicate the government's commitment for improving the sector and creating a positive environment for entrepreneurship development in the field.

Being a new concept, the requirements of scientific and commercial floriculture is not properly understood in the country. The developmental initiatives of the government have to keep in mind the low knowledge base, small land holdings, unorganised marketing and poor infrastructural support. While long experience of flower growing in the open field conditions enable sufficient flower production for domestic markets, the quality of the produce, in view of its exposure to various kinds of biotic and abiotic stresses, is not suitable for the ever growing export market. The production technology for flowers under protected environment of green houses needs to be standardised. There is hardly any post harvest management of flowers for the domestic market.

Availability of surplus flowers from exports for sale in the domestic market, has increased the appreciation of quality produce and the demand for good quality flowers is increasing. With the introduction of new varieties of crops in the country, facilities for generating their planting material for large scale production need strengthening. Special attention needs to be paid to strengthen the marketing infrastructure like organised marketing yards, auction platforms, controlled condition storage chambers etc. Greater research efforts are also needed for integrated pest management, development of location specific package of practices for traditional flowers, value addition to traditional flowers etc.

The initial cost and availability of finance is a critical matter in the development of large commercial projects requiring heavy investments. More options for developmental finance, such as the soft loan scheme of the National Horticulture Board need to be identified. In the initial years of commercial floriculture development, the governmental support in terms of subsidies etc. needs special attention.

The potential for growth of export market is always linked to the strength of domestic market - its capacity to absorb surplus and over production, and quality consciousness of consumers. Though we have a large domestic market, the marketing system and facilities need to be modernised.

The production for exports at present has suffered due to a few constraints. While our growers have been successful in producing world class quality at low cost, high air freight rates, low cargo capacity available, imposition of import duties, inadequate export infrastructure etc. have reduced their competitiveness. There is also a shortage of trained manpower to handle commercial floriculture activity. The demands of the growing export oriented industry would require adequate attention to be paid for human resource development, particularly at the supervisory level.

Table 2. Cut flower exports from India

Country	1993-94	1994-95	1995-96
Japan	322.50	8255.58	35932.56
Netherlands	1004.61	9102.49	24799.90
U.S.A.	1175.38	2495.21	17652.50
Germany	957.61	2538.63	9256.00
U.K.	1420.93	1113.78	3345.86
U.A.E.	2120.40	3388.19	2459.29
Italy	210.28	164.96	2200.49
Hongkong	730.02	903.15	1504.02
Singapore	78.45	437.37	1190.54
Nepal	11.86	36.09	292.64
Kuwait	24.22	5.00	274.13
Saudi Arabia	413.98	169.52	272.77
Switzerland	136.96	258.20	242.83
Hungary	—	286.69	181.82
Thailand	—	86.49	177.82
Australia	—	—	132.89
Russia	368.46	20.96	119.85
Others	988.68	719.39	9293.82
Total	9964.34	29981.60	109329.73

Marketing of Cut Flowers

Marketing of cut flowers is very unorganised at present. In most metropolitan cities, with large market potential, flowers are brought to wholesale markets, which mostly operate in open yards. A few large flower merchants generally buy most of the produce and distribute them to local retail outlets after significant mark up.

The retail florist shops also usually operate in the open on-road sides, with different flowers arranged in large buckets. In the metros, however, there are some good florist show rooms, where flowers are kept in controlled temperature conditions, with considerable attention to value added service. The government is now investing in setting up of auction platforms, as well as organised florist shops with better storage facilities to prolong shelf life.

The packaging and transportation of flowers from the production centres to the wholesale markets at present is very unscientific. The flowers, depending on the kind, are packed in old gunny bags, bamboo baskets, simple cartons or just wrapped in old newspapers and transported to markets by road, rail or by air. The mode of transportation depends on the distance to the markets and the volume. Mostly, flowers are harvested in the evening time and transported to nearby cities by overnight trains or buses.

In recent years, the government has provided some assistance for buying refrigerated carriage vans. A large number of export oriented units have built up excellent facilities of pre-cooling chambers, cold stores and reefer vans and their produce coming for domestic market sales are thus of very good quality and have longer vase life and command higher price. The government programmes for floriculture development include creating common facilities of cool chain in large production areas to be shared on cooperative basis. Formation of growers' cooperatives/associations are being encouraged.

In view of the unorganised set up, it is difficult to estimate the size of flower trade, both in terms of volume and value. A study conducted in 1989 estimated the trade to be worth Rs. 2050 million. It is in the period of the last five years or so that this business has really boomed in India, which is reflected in the number of new florist outlets in all cities and increase in the public's purchase of flowers as gifts. This would put the current trade at several times the earlier estimate. A recent study of Delhi market alone put the value of flowers traded on wholesale as Rs. 500 million.

The loose flowers (traditional crops like marigold, jasmine etc.) are usually traded by weight. The average price of different flowers in major markets varies considerably depending on the period of availability (Table 3).

Table 3. Average market price for major flower crops

Flowers	*Unit*	*Price (US$1 = Rs.40)* *Rs./kg or doz or each stem*
Marigold	kg.	3-60
Jasmine	kg.	15-150
Crossandra	kg.	20-120
Chrysanthemum	kg.	5-25
Tuberose	kg.	5-30
Rose	kg.	6-60
Gladiolus	doz.	20-75
Carnation	doz.	30-75
Gerbera	doz.	36-75
Orchids	each stem	10-45
Liliums	each stem	10-45
Anthuriums	each stem	15-45

The net returns to the growers depend on the packaging and transportation costs. The cut flowers with stem have a limited overall market in terms of volume. The share of cut flowers has almost doubled from 30 to 60% in the last decade.

The value of cut flower export from India has increased twenty five fold during the last five years (Table 4). With more export oriented units coming into operation, exports are likely to grow further in the coming years. The major share of the export trade is for roses, in addition to orchids, gladiolus etc. The major markets are Europe and Japan. The exports of roses to Japan, have really picked up in the three years from Rs. 360 million in 1993-94 to Rs. 6090 million in 1995-96. As per the estimates for 1996-97, India has been the largest supplier of roses to Japan (volume wise).

Table 4. Export of floriculture products from india (Rs. in Million)

Item	*1991-92*	*1992-93*	*1993-94*	*1994-95*	*1995-96*
Cut Flower Fresh	3.99	10.90	9.96	29.98	100.33
Dried Flowers	64.35	74.31	109.21	195.78	364.56
Live Plants	40.15	30.56	30.52	60.43	81.48
Dried Plants	23.99	25.45	23.72	10.35	35.83
Bulbs, Tubers etc.	12.06	7.83	14.95	11.83	19.21
Total	144.54	149.05	188.36	308.37	601.41

The internation cut flower market has changed dramatically over the years. The Dutch dominated the flower growing market in the '80s with new varieties. Competition from Third World countries increased as they took a lion's share of the "traditional" domestic cut flower market. Rising freight costs put a damper on shipping from one coast to the other.

The cut flower market is based on supply and demand, but you can often create your own demand by offering high quality and unusual product with reliable service to back it up. Crops produced in the cooler "shoulder" seasons can demand a higher price since the market is not typically flooded with product. Especially in reference to the small or starting grower, the following phrase is repeated over and over in both the research and industry literature: quality sells. Grow for quality, and don't be afraid to charge for it - price will follow quality.

The profitable grower does not wait until he or she has harvest-ready flowers to decide their fate. Have a clear market plan established ahead of time. Your target market influences what you will grow, how it will be handled and packaged, and most important, your capital investment.

Cut flowers usually are sold by the bunch, in arranged bouquets, or individually. Marketing options include selling to wholesalers, florists or other retail outlets, or direct to the public. The obvious route when first starting out is to target local markets. As your volume increases, you may want to deal with wholesalers and distributors.

"Direct to the public" sales include farmer's markets, roadside stands, or cut-it-yourself arrangements. How much you wish to interact with the public can help determine if a cut-your-own business is right for you. On-site sale, whether in the field or from a stand, requires a good location in a populated area, and a desire to deal with (not a mere tolerance of) the public. Production considerations may be altered a bit for the "cut your own" concept. If you are integrating a field cut flower business with a preexisting greenhouse or nursery that retails to the public, the additional marketing requirements would be minimal.

There are many direct-sales opportunities away from the farm as well, though transportation now becomes a cost factor. Farmer's markets offer a low-overhead market for novice growers. You can experiment with displays, mixed bouquets, dried materials, etc., and enjoy relatively autonomous pricing. There are many other retail opportunities available, such as restaurants, banks, etc. Craft shows are great sales venues for preserved and dried plant material.

Sales to retailers such as florists, garden centers, grocers, and upscale or gourmet specialty stores are other options. Industry experts suggest that for businesses other than florists, start with a few sizes of mixed bouquets, then move to quantities of single species. Offer pre-made bouquets as a labor-saving option to florists. Florists are constantly searching for new and unusual material.

Selling to wholesalers entails meeting their grading, sorting and packaging requirements, and consistency is greatly appreciated. Though price will be lower, wholesalers will often accept large quantities. Currently, the Internet offers a new marketing niche. One can provide product information in a website to potential customers, product availability lists updated weekly or even daily, and create interactive order forms. Regardless of your market, consistency of quality and service should be your business goals. In order to compete with the offshore market, local growers must offer comparable quality, grading, packaging, and promptness.

3

Production and Marketing of Bedding Plants

Bedding plants include a large number of annual and perennial plant species that provide seasonal colour in the landscape and in hanging baskets. In the past decade, no other segment of the floriculture production industry has enjoyed more public interest and use of its product than bedding plants. Bedding plants have become an indispensable item for landscape use, presenting an array of flowers and foliage that add colour and texture to the landscapes of homes, apartment complexes, shopping malls, public buildings, city streets and parks.

The largest consumer of bedding plants is the general public, the second largest is landscape companies. It is difficult to strictly define bedding plants because annuals and perennials are grown in the same greenhouses and are sold and shipped together. Plant types and production systems are numerous and diverse.

Bedding plants are sold to wholesale and local and national retail outlets and landscape companies. The market division varies tremendously among businesses. The larger businesses and a few of the small businesses primarily sell wholesale. Most of the small and medium-size businesses sell portions of the crop at wholesale and portions at retail. A few smaller greenhouse businesses primarily grow plants that are sold through their own retail outlet. Forty-eight commercial (wholesale/retail) greenhouse businesses belong to the Washington Floricultural Association and Puget Sound Flower Growers Association. Supply businesses, community colleges, and parks are also members of these associations but not counted in this profile as they are not commercial bedding and flowering plant producers.

Bedding-plant Production

Bedding-plant production is an important component of the floriculture industry. Each year, many small businesses spring up selling bedding plants by the roadside. The consumer's demand for plants in the spring seems enormous, and the industry continues to grow at a strong pace.

Many different kinds of businesses want to get some of the market and sell bedding plants. Bedding-plant sales often seem to be a sure thing, but finding a market before planting the first seedling or plug is the smartest business strategy for any bedding-plant grower. Start by planning to grow a small number of the most popular bedding plants.

In successive years, plan to increase the number gradually if sales were what was projected. If this is the first season growing bedding plants, check out the competition and the products they sold in the past year. Check the prices of the competition and make some estimates of how much it will cost to grow bedding plants. Many established growers in the area are often willing to share marketing information. Work out a plan on paper to produce plants and sell them at a profit.

Production Planning

Follow five steps each year to plan a bedding plant production schedule.

1. Decide who makes up your market.
2. Decide when to market your plants.
3. Select which kinds of plants and which cultivars of those plants to grow.
4. Determine how you will produce them.
5. Calculate at what price to sell them for a profit.

First, decide who makes up your market. Many kinds of consumers use bedding plants throughout the year. Homeowners are perhaps the largest market because they want a variety of bedding plants. Landscape contractors, landscape maintenance firms, garden centers, home centers, supermarkets, and discount stores also buy bedding plants. Talk with these potential customers several months to a year before they will need the plants. Many of these businesses will make a contract with a grower to produce a cer-tain number, size, and type of bedding plants well ahead of their need. They will also need some bedding plants only a few weeks before sales.

Second, decide when customers want to buy plants. Consumer demand for bedding plants is greatest in spring, yet there is also some demand for bedding plants in the summer months and in the fall. Determine when plants should be ready for sale, then plan from that market date back to a planting date. Bedding plants sell best when several flowers are in colour. Be sure to plan for enough time to produce flowers on the plants, especially if that's what customers specify.

Third, determine which plants and which cultivars of those plants to produce. Several factors should be considered when deciding which bedding plants to grow. Two prime considerations are popularity among consumers and performance in the garden, especially heat tolerance. You may not be able to grow everything that the customer wants. Some crops, such as begonias, are expensive to grow because they take a very long period of time to produce. Others are not heat tolerant. When shipping to out-of-state markets, find out which plants work well in particular climates.

What have homeowners been buying? Impatiens have been the best selling variety of bedding plant for 10 years in a row. Consider what people bought last year when planning for next year. Consumers' past purchases are a very good indication of what they will want to buy next year. Plan, too, for production of a few of the good-performing but unusual plants. The newer items often sell well because some people want plants just a little different from their neighbors' plants.

Selecting Heat Tolerant Varieties

Anticipate the stress that summer heat and drought will put on plants outdoors and consider growing those cultivars that perform well in the South. Customers may buy plants that are not heat tolerant this year because they are attractive, but they won't be back for more next year because the plants performed poorly. Several universities and botanical gardens conduct outdoor trials of new bedding plant cultivars and types for garden performance.

Cultivar Series

Cultivars of many bedding plant species are marketed by seed companies as groups called series. These are often breeding lines with similar genetic backgrounds.

Cultivars within a series often have similar characteristics such as plant size and shape, environmental tolerance, and cultural requirements. However, the individuals within a series have different flower colours and different colour patterns.

Selecting a variety of colours from the same series allows the grower to provide the market with a variety of colours to choose from while keeping cultural requirements simple.

Production Structures

A greenhouse structure with a heat source is usually required; however, crops with short production times may be produced outside, in cold frames, or in greenhouses without a heat source. A greenhouse constructed of double-layer plastic, with the plastic layers 2 to 4 inches apart, can make a lightweight structure ideal for bedding-plant production. The plastic is stretched over an aluminum frame to form a Quonset structure. The ends of the house should have plastic covers that can be easily removed when the temperature climbs in the spring. Ventilation to bring cool air inside on warm days is just as important as heating on cool days.

Propagation Schemes

The majority of bedding plants are propagated from seed, usually sown from late December into March. Ten to 15 years ago most bedding plant seeds were sown and germinated in open containers, usually wooden flats, or sown directly in the final container. Once the seedlings were large enough to handle (usually 2 mature leaves), they were pricked out by hand and transplanted to market flats for finishing. Therefore, most of the production time was spent in the final container. However, this method had several problems:

1) a lot of time and labor was required for transplanting,
2) diseases could spread easily from seedling to seedling in an open tray, and
3) damage to sensitive roots caused transplant shock.

Seedlings were essentially uprooted in the transplanting process and had to start over developing a root system.

The development of plug technology in the 1970s and 1980s resulted from efforts to overcome the problems associated with the traditional method. As the name suggests, these little plants are ready to be plugged into a pot or cell pack to produce a mature plant in an average of 3 to 5 weeks less time than it takes to produce the crop from seed.

A plug is the individual cell in a plug flat. A plug flat is a plastic waffle-like tray (11" x 22" in size) composed of 50 to 800 individual cells, each 2- to 5-cm deep. Plug flats are named and sold based on the number of cells available. Bedding-plant seeds are sown directly in plug flats filled with media by a mechanical seeder, germinated in the flats, and grown to a transplantable stage. Plug technology has the following advantages:

1) Reduces the time and labor required for transplanting because it is easier and faster to remove one seedling as a unit
2) Maximises greenhouse space utilisation because seedlings are perfectly spaced and remain in the plug tray longer
3) Increases seedling and final crop uniformity because faster growing seedlings in a plug tray crowd early and growth slows while slower seedlings have a chance to catch up

4) Reduces transplant shock because roots are not disturbed and the seedling continues growth as soon as it is transplanted
5) Reduces the spread of disease organisms because each seedling is isolated
6) Reduces crop duration because no transplant shock occurs
7) Mechanises almost every phase of plug production
8) Provides an easier and less expensive way to ship seedlings than traditional wooden trays provide

The main disadvantages of the plug system are that the initial cost for equipment such as automatic seeders, flat fillers, and environmental control facilities for germination is high, and extreme skill and careful attention to detail are required for successful plug production. Beyond the large differences between the traditional and plug methods during germination and initial seedling stages, finishing techniques in the final containers are essentially the same.The majority of bedding-plant growers utilise plugs for finish production. However, a major production decision is whether to buy plugs from specialist plug growers or to invest in the equipment and skill needed to grow plugs in-house. Much of this decision involves comparing the cost of buying plugs versus producing plugs. The advantage of buying plugs, especially for the new grower, is the elimination of the germination and early seedling growth stages, which can be costly and have greater risk of crop loss.

Many small- to medium-size growers essentially choose to be bedding-plant finishers, buying plugs from specialist plug growers and dealing only with transplanting and finishing. This avoids the cost of equipment, cost of skilled plug growers, germination problems, and other problems associated with the seedling stage. Specialising only in finishing crops results in the ability to tightly schedule the production area for rapid crop turnover.

Many larger growers may grow plugs not only for their own production but also for sale to other growers. In-house plug production is done to reduce the cost per seedling unit, control plug quality, and grow cultivars that can be difficult to find from specialist plug growers. Often, sales of plug flats to other growers is necessary to justify the cost of equipment and skill required for plugs.

Plug Production

Seed

Producing plug-grown bedding plants has resulted in a demand for high-quality seeds that germinate uniformly. Clearly, plug growers desire a high germination rate and seedling survival not only for in-house production but for customers as well. The combination of mechanical seeders and market demand has placed pressure on seed companies to consider seed quality as part of the overall effort to breed new cultivars. Several seed treatments are used in an effort to improve seedling stands in plug flats. These are the following:

Refined Seed

These are cleaned seeds that have been physically separated by size, shape, weight, or density. Typically, individual seed with poor germination and survival may be smaller than normal, have a different shape, or weigh less than normal. Mechanical techniques have been developed to remove unusual seeds from a lot of seed.

Primed seed

These seeds have been soaked in a solution with a high osmotic potential so that seed water uptake proceeds to a certain point, then stops. Thus, pregermination metabolic activities proceed but emergence of the radicle does not occur. The seeds are then removed for the solution, dried, packaged, and labeled with a sow-by date. The main benefit of primed seed is that germination is more uniform over a wider range of environmental conditions.

Seed pellets

These small seeds can be covered with a thick sugar coating to facilitate sowing by a mechanical seeder. The coating may be coloured for visibility on the plug medium.

Plug Medium

A medium for plug flats, like other container media, must have good aeration and be well drained. However, because of the small soil volume of each cell, the medium components must have a small particle size. Peat, vermiculite, and perlite are the most commonly used components, but finer grades are employed. In addition, a plug medium must have a high water holding capacity and a high ability to hold fertilizer because of the small soil volume and the fact that seedlings may be grown in the medium for 4 to 8 weeks. Like many container mixes, medium pH for plug mixes is adjusted using dolomitic limestone, and superphosphate and micronutrients are added, though at much lower rates.

Little else is added to a plug medium because soluble salts levels considered low for many pot plants are too high and may inhibit germination of seed. Many commercial companies produce special bagged mixes designed for plugs.

Seed Germination

Generally, temperature, oxygen, light, and moisture must be controlled for optimum seed germination. Temperature is probably the most important factor regulating the rate of seed germination. Germination rate is invariably low when the temperature is lower than optimum for a given species. As temperature rises, germination rate increases and remains steady within a range optimum for the species. Above optimum, the rate declines again. .

Excessive moisture or poorly drained media can reduce the oxygen available to seed and reduce germination percentage. Most bedding-plant seed have a relatively high humidity requirement during the time of water uptake, usually in the 90 to 100 percent range. However, constantly covering the seed in a thick film of water can limit the oxygen supply to seed for many species and inhibit germination. The seed of some bedding-plant species require light for germination so these should be sown on the medium surface without seed covering. Others bene fit from some covering, usually with sifted vermiculite, either because germination is inhibited by light or light plays no role and germination is enhanced by the extra moisture.

Plug Stages

Germination and growth of seedlings in plug flats have been divided into four easily recognizable stages. These stages are useful because environmental and cultural requirements for the plant changes each stage. The stages are as follows:

— *Stage 1*: From sowing to radicle (root) emergence. Generally, warm temperatures and high humidity are required. Nutritional requirements are very low and high soluble salts can be damaging. However, a few species may benefit from 15 to 50 ppm from KNO_3. Low oxygen can inhibit germination.

— *Stage* 2: From radicle emergence to emergence of the first true leaf. The goal is to develop a strong root system and prevent the seedling from stretching. Reduce temperature and humidity while providing adequate light.

— *Stage* 3: From the first true leaf to 4 or 5 leaves. Generally, the seedlings are ready to transplant at the end of this stage. Fertilise at 150 to 200 ppm N during this stage. Usually, a transplantable stage is reached when leaves of adjacent seedlings in a plug tray begin to touch and the root system has developed to a point that the seedlings can be pulled from the trays with the medium and roots intact.

— *Stage* 4: From 4 to 5 leaves (transplant), through flower formation to toning the plant in preparation for shipping.

Growth Retardants

Growth retardants are often applied to pluggrown seedlings to decrease internode elongation and to strengthen the stems. Though many label directions recommend application 2 to 3 weeks after sowing, it's best to apply growth retardants at the correct stage of development. For most species, this is at the second true leaf stage. Growth retardants are generally applied at a reduced rate compared to larger plants and, if required, reapplied as needed. Seedlings are more susceptible to phytotoxicity and stunting from incorrect application of growth retardants. The most effective application temperatures are 55 to 70 degrees F. Therefore, reduce the rate and apply only under cool conditions.

Germination Facilities

Facilities for seed germination and early growth of seedlings in plugs must provide for more exacting environmental control than is normally found in most greenhouses. The germination area must provide a high degree of control over temperature, light, and humidity; usually two or three different environments must be established to meet the needs of a wide variety of bedding-plant species. In a greenhouse, this can be accomplished by using several small greenhouse sections with above average heating, cooling, and sophisticated environmental control systems.

The best way to provide a high degree of humidity control and avoid excessive moisture around the seed is to reduce the moisture droplet size far below that which is usually employed by conventional mist used for propagation of vegetative cuttings. Fog systems have been developed that work far better than conventional mist and convert 90 percent of the water droplet size to between 5 and 20 microns. When a system like this is employed, ventilation must be restricted and carefully controlled to keep the environment even.

Beyond the greenhouse, many plug growers have developed and use specialised germination rooms. These rooms are constructed with treated lumber and heavy insulation and are moisture-tight on the inside to reduce temperature fluctuations. Ultra-fine fog systems are used to achieve a high degree of control over germination humidity. Light is provided from fluorescent lamps that are turned on for 16 hours per day. Air conditioners may be used in those rooms designed for cool-season crops such as pansy. Plug flats are placed on shelves spaced 2 feet apart. Seeds generally only remain in the germination room to the end of stage 1; any longer and the seedlings may stretch from insufficient light.

Finishing

Growing-on or finishing refers to the stage (stage 4) of production from transplanting into the final container until the crop reaches a marketable stage.

Bedding Plant Containers

Most bedding plants are produced and sold in market flats. A market flat is composed of a sheet of market packs (called an insert) inserted in a plastic flat. An insert is a thermo-formed, polystyrene plastic sheet of pull-apart packs. The packs are composed of 1 to 9 cells, each containing a single plant, designed for customers to pull-apart from the insert to purchase as a unit. The insert rests in a plastic flat to provide support and is the single unit for shipping. The choice of which container type and size to use for bedding plants is largely determined by the ultimate finish size of the plant species and cultivar grown, market demands, and constraints imposed by mechanisation.

Each species has a characteristic height and spread that influences the practical container size to use. Usually, a given species may be grown in a range of sizes and, within this range, finishing times will vary. But if the container size is too small, plants will crowd and lose lower foliage before flowering. Conversely, if the container size is too large, the plant will flower before filling the container to an aesthetically acceptable extent.

Probably, a far more important factor is to meet the demands of the market to which the product is sold and make a profit. The entry of mass marketing has had a profound effect on the choice of container sizes and, therefore, on the growers themselves. Rather than leaving the choice of container size to the grower, mass markets have sought to impose container size standards that fit within their marketing strategies. Because of customer specification and tight competition, the choice of containers is often very limited. This usually means smaller container sizes, a shorter finishing time, and a tighter profit margin.

Growers who market to traditional retail outlets, however, strive to choose container sizes that differentiate the product from the competition in terms of quality and appearance. The key is for customers to be able to identify your produce by appearance and good garden performance. Larger soil volume combined with quality growing techniques should produce a larger, more vigorous plant that is ready to provide quick gratification in the home garden or commercial landscape.

Given a choice, select the largest feasible market pack to accommodate the root system of the bedding plants to be grown. Larger market packs mean fewer plants per flat, and since most plants are priced by the flat, the number of cells per flat will affect costs and profits. Inserts come in 24, 36, 48, or 56 cells to the flat. Plants grown in a larger-sized cell will generally last longer in the sales area than plants grown in small-sized cells.

Standard Market Flats

The 11- by 21-inch plastic flat is still the standard for the industry. Inserts come in a wide variety of combinations of packs per flat and cells per pack. In catalogs, inserts are often designated by a number. The first part of the number indicates the number of packs and the last part, the number of cells per pack, (e.g., a 1204 insert has 12 packs each with 4 cells for a total of 48 plants in a flat). The most widely used flat is the 12 packs per flat, each pack with 2, 3, 4, or 6 cells per pack. Mass markets have popularised the 18 packs per flat with 1 to 4 cells per pack, while growers for garden centers often use the 8 packs per flat with 4, 6, or 9 cells per pack.

Jumbo Flats

One of the main growing areas of the flat market is the jumbo pack that has 6 packs per flat with 4 or 6 cells per flat. The flat size remains the same as standard flats, but the cells are 31/8 inches deep instead of the 23/8 inches for standard packs providing a 25 to 30 percent increase in soil volume. Jumbo packs are used by growers seeking longer shelf-life, reduced losses in the retail area, and better quality for the customer.

Slimline Flats

For the large grower who markets to chain outlets, slimline flats offer a way to reduce production costs when the market resists a price increase. These flats are 81/2 inches wide (rather than 11 inches) by 20 or 21 inches long. Therefore, they take up 15 percent less space than standard flats and more flats can be grown per unit greenhouse area.

Bedding plants marketed in containers larger than market flats are becoming more popular each year for both homeowners and the landscape contractor markets. Some bedding plants grown in 4- or 6-inch pots will produce a colourful show in several weeks and bring a premium price in the market. Even larger-sized colour pots, containers with multiple plants of different species, or cultivars mixed in a colourful display, are becoming more popular. Showing flower colour is essential to marketing these larger products. Start growing larger-sized products by transplanting market-pack-grown plants to large containers early in the spring.

Growing Media

Soilless mixes are the most often used growing media for bedding plants. These mixes provide anchorage that enables the plant to support itself and regulate the supply of water, oxygen, and nutrients to the roots. No one specific medium can be singled out as the best, and quality bedding plants can be grown in a wide variety of media combinations. The type of medium a grower chooses depends on personal preference, type of greenhouse, type of irrigation, crop type, and cost. Some soilless mixes work better for some types of plants than do others, but many growers use one mix for all bedding plants.

The first decision involving growing media is often whether to purchase medium or mix your own at the greenhouse operation. On face value, commercially bagged mixes seem expensive but once all the costs are considered, mixing your own formulation may be out of the question for most small growers. Like many economic decisions in the greenhouse, carefully consider all the costs of both approaches. For a commercial bag mix consider the cost per bag and shipping costs. For mixing your own formulation, consider each component's cost, management time, labor, office expenses, equipment costs and depreciation, and costs of special structures for the mixing facility and component storage. Many large growers mix their own formulation because the cost of commercially bagged mixes would be prohibitive. One other option is to purchase bulk mixes from local formulators. These companies will mix a medium to your specification and deliver by truck at a lower cost than you would pay for commercially bagged mixes.

Regardless, make sure that the medium comes from a reliable source so that there are few weed seeds and potential for insect or disease problems. Fertiliser and other components should be well blended to ensure a medium that is uniform throughout the batch and from bag to bag. The media should be about half peat moss and about half of some aggregate, such as perlite or vermiculite. This will help the media to hold water yet allow the excess water to drain adequately.

The pH of the media should be between 5.5 and 6.5 to ensure that the nutrients you apply in fertilising can be absorbed by the plants. A pH outside this range will tie-up the nutrition applied in fertiliser, the plants will become deficient in those nutrients, and you will have plants with yellow or brown leaves. Some elements become toxic to the plant at a pH below 5.0.

Selection and Use of Stress-Tolerant Bedding Plants

Each of us are subjected to stresses and pressures every day in our home, work, and living environment; plants are no different. Unfortunately, there is no "stressless" environment, and there is no totally stress-resistant bedding plant. Each site has its stress level and each plant has its tolerance level. There are steps that can be taken to reduce or avoid stress in the landscape. However, no programme can prevent all problems, and the key to successful landscape colour using bedding plants is to match the particular site with specific plant species. But before you can select plants to use, the site should be accurately analysed and characterised, and preparations should be made to minimise stress conditions that may occur.

Site Analysis

A site analysis for bedding plants should include (1) temperature averages for the colour season, (2) amount of sunlight received daily, (3) rainfall averages and average intervals between rains, and (4) soil characteristics such as drainage and moisture retention. Each of these components should be further defined prior to plant selection.

Temperature. Very few species look attractive and flower profusely from early spring through late fall, and rotational planting for continuous colour should be considered (Table 1). Cool-season flowers such as dianthus, snapdragons, and pansies can be used early in the season. It is possible to extend the flowering season of cool-season annuals by placing them in a protected location, shaded from direct sunlight from 12:00 to 4:00 PM. Heatloving flowers such as gaillardias, portulaca, sand verbena, and vinca do not begin to flower until early summer and should be used for summer colour and high temperature situations. Heat tolerance is an advantageous characteristic, and bedding plant trials can offer cultivar suggestions to landscapers interested in plant performance in high temperatures (Table 2). Another temperature consideration is frost tolerance (Table 1). Avoid early planting of tender plants to prevent frost damage. Tender species also will be the first to be killed from frosts in the fall.

Light: Light and temperature are closely related, and plants listed as preferring lower light may tolerate more sun, if temperatures are moderate. When evaluating light exposure, note the duration and intensity of light the site receives. Four hours of full sun during the morning is much different than four hours of afternoon sun. Also, in a shaded location, the degree of light filtration can vary. In general, if the site receives more than 3 hours of unfiltered mid-day sun, it should be treated as a "full sun" site, with respect to plant selection (Table 1).

Table 1. Bloom period, frost tolerance, light preference, and relative drought tolerance of bedding plants.

Name	Bloom Period	Frost tolerance	Light	Drought tolerance
Ageratum *Ageratum*	Summer	Tender	Full sun	Moderate
Alyssum *Lobularia*	Spring-early fall	Tolerant	Full sun to partial shade	Low
Aster *Callistephus*	Summer-early fall	Moderate tolerance	Full sun to partial shade	Low
Begonia *Begonia*	Late spring-early fall	Tender	Full sun to heavy shade	Low
Blanket Flower *Gaillardia*	Summer-early fall	Tolerant	Full sun	Moderate
Browallia *Browallia*	Summer	Moderate tolerance	Full sun to partial shade	Low
Calliopsis *Coreopsis*	Summer	Moderate tolerance	Full sun to partial shade	Moderate
Candytuft *Iberis*	Late spring-early summer	Tolerant	Full sun to partial shade	Low
Calendula *Calendula*	Summer-	Tolerant	Full sun to fall partial shade	Low
Celosia *Celosia*	Summer-early fall	Tender	Full sun	Moderate
Coleus *Solenestenon*	Late spring-fall (foliage)	Moderate tolerance	Partial to heavy shade	Low
Cornflower	Summer	Moderate tolerance	Full sun to	Low

(Contd...)

Name	Bloom Period	Frost tolerance	Light	Drought tolerance
Centaurea			partial shade	
Cosmos *Cosmos*	Spring- early fall	Moderate tolerance	Full sun	Moderate
Dahlberg *Dyssodia*	Daisy Summer- fall	Moderate tolerance	Full sun	High
Dahlia *Dahlia*	Summer- fall	Tender	Full sun	Low
Dianthus *Dianthus*	Spring and fall	Very tolerant	Full sun to partial shade	Low
Dusty Miller *Senecio*	Spring- fall (foliage)	Tolerant	Full sun to partial shade	Moderate
Geranium *Pelargonium*	Late spring- early fall	Tender	Full sun	Low
Globe Amaranth *Gomphrena*	Late spring- early fall	Moderate tolerance	Full sun	High
Gloriosa Daisy *Rudbeckia*	Summer	Tolerant	Full sun to partial shade	Moderate
Hypoestes *Hypoestes*	Summer (foliage)	Tender	Partial to heavy shade	Low
Impatiens *Impatiens*	Late spring– early fall	Tender	Full sun to heavy shade	Low
Lisianthus	Mid-summer to	Tender	Full sun	High

"Partial shade" can be defined as receiving unfiltered morning sun, but shade during the afternoon hours, or moderate shading throughout the entire day. A "heavily shaded" site would receive very little direct mid-day light and less than 60% of the sun's intensity during the remainder of the day. A mismatch of plant and light can lead to reduced flowering, leggy growth habit, burning of plants, and stunting of growth.

Water: Water stress covers both extremities of the spectrum, even for the same landscape site. Bed preparation is essential for avoiding both moisture excess and drought conditions. For most situations, supplemental irrigation will be required at some point during the growing season. For minimal irrigation sites, select "drought tolerant" species. The best insurance against excessive moisture is proper bed preparation and sufficient drainage. Keep in mind that the majority of overwatering problems, assuming a well-prepared site, occur from too frequent irrigations rather than too much water applied at any one time. If supplemental irrigation is in place, apply enough water at every watering to assure complete bed coverage. Also, an irrigation schedule should take into account rainfall and be adjusted appropriately to be most effective.

Soil Characteristics: Plants depend on the soil for water, anchorage, and nutrients. Frequent heavy rains in combination with poorly drained beds will reduce plant performance and increase the chances of root rot problems. On the other hand, beds with excellent drainage combined with little water holding capacity could require irrigation as frequently as every other day. Nutrient deficiencies and toxicities are common in the landscape, although they are easily avoided if proper steps are taken. Do not guess at fertility levels-take a soil test and send it in for analysis.

Table 2. Heat and humidity tolerant bedding plants.

Name	*Colour*
	Bronze-leaved Begonia
'Brandy'	Pink
'Whisky'	White
'Espresso White'	White
'Vision'	Red
'Bingo Red' Red	
'Espresso Rose'	Rose
'Bingo Rose'	Rose
	Green-leaved Begonia
'Prelude Pink'	Pink
'Viva'	White
'Prelude White'	White
'Varsity Scarlet'	Red
'Scarlanda'	Red
'Eliza'	Rose/Salmon
'Ambra'	Rose/Salmon
'Rum'	Bicolour
	Geranium
'Pink Orbit'	Pink
'White Orbit'	White
'Ringo White'	White
'Pinto Red'	Red
'Pinto Rose'	Rose
'Hollywood Rose Pink'	Rose
'Pinto Salmon'	Salmon
'Hollywood Star'	Bicolour
	Impatiens
'Dazzler White'	White
'Impulse White'	White
'Impact Rose'	Rose
'Novette Deep Rose'	Rose
'Dazzler Coral'	Coral/Salmon
'Impact Coral'	Coral/Salmon
'Impulse Salmon Orange'	Coral/Salmon

Contd....

Contd....

'Accent Rose Star'	Star
'Spotlight Mix'	Mix
'Impulse Carmine'	Carmine
'Impact Carmine Rose'	Carmine
'Accent Deep Pink'	Pink
	Impatiens
'Dazzler Blush'	Blush
'Accent Bright Eyes'	Blush
'Super Elfin Blue Pearl'	Lilac
'Accent Lilac'	Lilac
'Dazzler Burgundy'	Burgundy/Purple
'Super Elfin Violet'	Burgundy/Purple
	African Marigold
'Perfection Gold'	Gold
'Perfection Orange'	Orange
'Voyager'	Yellow
	French Marigold
'Red Marietta'	Single Red
'Orange Boy'	Double Orange
'Bonanza Orange'	Double Orange
'Bounty Gold'	Double Gold
'Early Queen Sophia'	Double Bicolour
'Hero Harmony'	Double Bicolour
	Petunia
'Pink Carpet'	Pink
'Eterna Pure Pink'	Pink
'Pink Madness'	Pink
'White Carpet'	White
'Celebrity White'	White
'Eterna Vivid Red'	Red
'Rose Madness'	Rose
'Rose Carpet'	Rose
'Deep Rose Pearls'	Rose
'Coral Madness'	Coral
	'Velvet Picotee
Improved'	
Bicolour	

Contd....

Contd....

'Electra Blue'	Blue
'Eterna Lilac'	Burgundy
'Supercascade Lilac'	Burgundy
'Purple Pirouette'	Double Purple
	Vinca
'Peppermint Cooler'	Upright White/Red Eye
'Grape Cooler'	Upright Pink/Rose Eye
'Rose Carpet'	Prostrate Rose

Air pollutants: Some landscape sites, especially in highly urbanised areas, are subjected to significant levels of air pollution. The most damaging of these pollutants are sulfur dioxide SO_2), ozone (O_3), and peroxyacetyl nitrate PAN). Symptoms of SO_2 injury include necrotic dead) spots between the major veins, where the tissue turns light tan and papery in texture. The most common symptom of exposure to O_3 is the formation of tiny, light-coloured flecks or spots on the upper surfaces of affected leaves, similar to spider mite damage. PAN injury is expressed as silvering, glazing, bronzing, and sometimes death of the lower leaf surfaces. Bedding plants do exhibit relative sensitivity and tolerance to these materials, and if pollutants are a problem, plants should be selected accordingly.

Transplanting, Production Temperature, and Spacing

Plug technology has enabled growers to automate even the transplant stage of bedding-plant production. However, most growers still transplant by hand today. Transplant seedlings into slightly moist media. The media should fill the cells of the market packs to but not over the rim. Seedlings should be inserted into the new medium at about the same level they were growing in the plug tray. Immediately after transplanting, the plants should be watered thoroughly and moved to the greenhouse. As a preventive measure, some plants may be watered with a drench of fungicide to prevent damping-off diseases. Root zone heating will help the seedling to establish a good root system in the new container. Monitor the temperature of the environment, which should be maintained at 75 to 80 degrees F day and 65 to 68 degrees F night temperatures. Water and fertilise when necessary, allowing the top of the soil media to dry between watering. Spacing of plants in market flats is determined by how many cells fit into a flat. Spacing can be regulated on larger-sized pots, such as 4- or 6-inch pots or hanging baskets. Plants can be grown pot to pot and spaced slightly apart in the last 2 to 3 weeks of production.

Fertilisers

Moderation is the key to successful bedding-plant fertilisation. Bedding plants should be provided all 16 essential elements either from the liquid fertilisation programme and/or incorporation at the time the medium is mixed. Most growers use a combination of both. Water-soluble fertilisers come in a wide range of formulation. Many growers successfully use a 20-10-20 formulation; however, 20-10-20 is an acidreaction fertiliser that can reduce the medium pH over time. If the pH of the media is too low, apply a basic-reaction fertiliser such as 15-0-15. Many growers will alternate between these two fertiliser formulations to maintain pH. Fertilisers especially for soil-less mixes should have a good supply of micro-nutrients and be low in ammoniacal nitrogen, a cause of ammonium toxicity during cool weather. As an average, 150 to 200 parts per million (ppm) nitrogen in the irrigation water works well. Reduce the concentration and frequency of fertilisation at the beginning and end of the production cycle.

Growth Retardants

Several bedding plants may begin to stretch, developing long, thin, spindly stems a few weeks after transplanting, especially under low-light conditions in early spring. To keep them shorter, sturdier, and greener some growers apply a growth retardant. Several growth retardants are approved for use on bedding plants. Growth retardants decrease the length of the stem (shorten the internode between leaves), thus developing a shorter, sturdier plant. Some growth retardants have an added benefit of greening up the foliage. Ancymidol (A-Rest) can be used as a foliar spray applied 2 to 4 weeks after transplanting at a rate of 30 to 130 ppm. Diaminozide (B-Nine SP) can be applied as a foliar spray 2 to 4 weeks after transplanting at a rate of 2,500 to 5,000 ppm. Bonzi, Cycocel, and Sumagic are also labeled for application to bedding plants. Plants differ in their sensitivity to these chemicals, so be sure to check the label before applying them.

Because no chemical growth retardants are approved for application to vegetable bedding plants (or fruit such as strawberry), the only chemical alternative is water, which is a powerful growth retardant. When plants are produced with sufficient but limited water, the result is a shorter, sturdier plant. Withholding excessive amounts of water and fertiliser, combined with sufficient sunlight and cooler temperatures, helps to reduce plant size and increase the quality of vegetable bedding plants.

Scheduling

Scheduling bedding plant crops is essential to greenhouse profitability. While quality and reasonable price are important, crops also must be ready when the market demands. Demand in the market, climate, geographic location, season of the year, and bedding plant type/cultivar all influence the dates for sowing, transplanting, and flowering. Despite the availability of a lot of information on bedding plant scheduling, careful crop records detailing finishing times over a period of years for a particular location are the best guide. Variations in microclimate within a city, even a particular greenhouse can make a difference of a week or so.

Scheduling bedding plant crops is often done by assigning the weeks of the year week numbers. The first full week in January is numbered week 1, then each week is numbered consecutively through December. Schedule a crop as follows:

1. Determine the week number of the sales date.
2. Count back the number of weeks required in the finish container. This is the transplant week number.
3. Count back the number of weeks required in the plug flat. This is the sowing week number.

Nutrient Deficiencies of Bedding Plants

Bedding plants are profitable crops to grow because of their rapid growth rate and popularity among consumers. Unfortunately nutrient deficiencies can appear quickly within a crop and subsequently reduce profit by affecting marketability. There are several factors which can inhibit proper nutrition of bedding plant crops. pH. One factor which can introduce nutrient deficiencies in bedding plants is the substrate pH. The general pH range for bedding plants is 5.4 to 6.8, but maintaining the pH between 5.6 to 6.2 is recommended. Poor uptake of nutrients, particularly boron, copper, iron, manganese, and zinc can occur if the pH of a substrate is above 6.5.

Certain macronutrients like calcium and magnesium can become less available at pH values below 5.4.

Improperly working equipment: An improperly working fertiliser proportioner can cause nutrients to be less than optimum in the substrate. Weekly calibration of the injector is required. The significant problem with equipment failure is the introduction of multiple nutrient deficiencies.

Water stress (too much): Constant saturation of the substrate can lead to macro and micronutrient deficiencies. As oxygen levels are inhibited by overwatering, root growth can be limited and water uptake slowed. Elements such as calcium are transported via waterflow and deficiency symptoms can develop rapidly on new growth. Also the inactivity of root systems due to saturated conditions can lead to inefficient uptake of iron or phosphorus.

Low soluble salts: Soluble salts refer to the total dissolved salts in the root substrate at any given time and is measured in terms of electrical conductivity (EC). When the EC content of the root substrate is too low, plant growth is stunted and mineral deficiencies are observed. Low salts are usually due to too many clear water irrigations. Deficiencies among bedding plants like lower leaf yellowing (nitrogen), lower leaf purpling (phosphorus), and lower leaf interveinal chlorosis (magnesium) are common when values are below 0.75 mS/cm (SME extraction).

Mineral antagonisms. When certain elements are provided in excess to plants, uptake of other nutrients may be hindered. One example of a mineral antagonism is the nitrogen-potassium interaction, where for most bedding plants a 1N:1K ratio is recommended. Other types of antagonisms are the potassium-calcium-magnesium interaction. Any one of these elements in excess can cause a decrease in the uptake of the other, therefore a ratio of 4K:2Ca:1Mg should be adopted by bedding plant growers. Excess phosphorus can cause a decrease in uptake of zinc, iron, and copper.

Temperature (too cold): Temperature can also play a role in the introduction of nutrient deficiencies. One classic example is the effect of low temperature (<55°F) on the uptake of phosphorus in tomato. Purpling of the lower foliage is the common symptom. Geraniums also can express phosphorus deficiency when they are grown too cool in the spring. Disease. Organisms like Pythium feed on the nutrients in roots causing an inefficient uptake of minerals.

Iron deficiency (upper foliage interveinal chlorosis) can occur if root rot pathogens infect the root system. Foliar diseases, particularly fungal diseases can cause chlorosis of leaf tissue, a direct reflection of harvesting nitrogen from plant cells.

Essential Nutrients

Sixteen elements are considered to be essential elements for plant growth: carbon, hydrogen, oxygen, nitrogen, phosphorus, potassium, calcium, magnesium, sulfur, iron, manganese, copper, boron, zinc, molybdenum, and chlorine. These elements have been determined to be essential because they have met these three criteria:

— The lack of the element makes it impossible for the plant to complete the vegetative or reproductive stage of life.

— The element cannot be replaced by supplying another element.

— The element must exert its effect directly on growth or metabolism.

Macronutrient/ Micronutrient Translocation

Knowing why the deficiency occurred is a crucial tool in the identification of a nutrient disorder. Another important aspect in diagnosis is the location on the plant where the symptom is expressed. Understanding the translocation principle in bedding plants will enable growers to diagnose more correctly, and will, in most cases pinpoint the macro or micronutrient disorder.

Mobility of the nutrient or the ability of the element to translocate itself to another part of the plant has been determined for the essential elements. Plants obtain nutrients from the substrate solution via root systems. Nutrients are incorporated into tissues, are used for cellular growth, and utilised in photosynthesis or in the building blocks of plant tissue.

Initially nutrients are provided to plants from seed reserves, then as roots begin to develop, plants extract nutrients from the substrate. Once the nutrients become less available to the plant, the plant has to provide nutrients from older tissues so that the actively growing regions (shoot tips and axillary shoots) can continue to develop. Basically, the plant is attempting to promote life by supplying nutrients to the upper growth, which harbours the reproductive structures, enabling the continuation of the species. Some elements are unable to be translocated. Differences between mineral translocation will be demonstrated below in a comparison example between nitrogen and calcium.

Nitrogen Translocation

Nitrogen is incorporated into organic molecules and is

involved in the structures of all amino acids, proteins, and many enzymes. As levels of nitrogen become less in the substrate, nitrogen is translocated from the lowest most leaves to the actively growing regions of the shoot tip. Deficiency symptoms appear on the older leaves and a lighter green colour is observed. As symptoms progress the stem becomes weak, the leaves become small and the lower leaves drop. Necrosis of the older leaves is an advanced nitrogen deficiency symptom.

Calcium Translocation

Calcium plays a major role in cell elongation and is an important component in cell walls, structurally it acts as a cement between cells. Calcium is transported with water to plant tissues, but if levels in the substrate are too low, calcium deficiency can occur. Because calcium is immobile, it cannot be translocated to the region of active growth in the shoot tip, therefore new growth is severely reduced. Although calcium may be adequate in the lowest most leaves, levels in the meristematic region can be too low, causing poor leaf expansion followed by necrotic patches in the young leaves. Complete necrosis of the shoot is the advanced stage causing the inability of the reproductive structures to form. If flowers were present when calcium levels become devastatingly low in the substrate, bud abortion occurs.

Fertilisation Strategy

Bedding plant producers will normally apply a commercial fertiliser to the crop. Normal fertilisation should provide adequate levels of micronutrients as well as the major macronutrients, as long as the pH is within the acceptable range.

Pests and Disease Problems

It is important to regularly check crops for insect and disease problems. Problems identified early are easier and more economical to control. Several precautions can be taken to reduce the likelihood of insect problems. Before using any insecticide or miticide, read the label directions carefully to be sure it is safe to use on specifiic crops and to double-check the recommended rate.

Eliminate weeds under the benches and around greenhouses to eliminate a major source of insect pests. Weeds are a great hiding and feeding place for many insects, especially whiteflies and aphids, before infesting crop plants. These insects will also feed on weeds while crop plants are being sprayed with insecticide only to return to crop plants later to feast. Several herbicides are registered to control weeds around greenhouses. Diquat liquid herbicide is registered for use under greenhouse benches for many types of weeds. Roundup (glyphosate) and Surflan liquid herbicides are registered for use as a spray to control several species of weeds growing under greenhouse benches.

Disease Control

One of the greatest disease problems, root rot, is caused by a group of fungi. Symptoms of this kind of disease are a general lack of vigor or growth, slow growth, and especially black roots or a blackening of the crown. These symptoms can be caused by any one or more of these fungi: Pythium, Phytophthora, Rhizoctonia, or Thielaviopsis. It is important for growers to watch crops carefully, especially at the earliest stages for these symptoms. Different fungi causing rot problems require different treatments. It is essential that the disease be diagnosed by a plant disease clinic in order to determine the proper treatment. Experience from treating past outbreaks will help you more quickly recognise and treat new occurrences. Refer to the product label for the correct amount to apply and read the label directions carefully to be sure that it is safe to apply the chemical to a specifiic crop.

Banrot 40W can be used as a soil or media drench at seeding or transplanting time as a preventive application for many bedding plants. It can help to prevent Pythium, Phytophthora, Rhizoctonia, Fusarium, and Thielaviopsis root rots. Truban 3G or Terrazole 5G (etridiazole) or Banrot 8G (etridiazole) can be used as a dry soil mix as a preventive measure against Pythium and Phytophthora for many ornamental bedding plants.

Subdue 2E can be applied as a soil drench to many ornamental bedding plants as a treatment for Pythium and Phytophthora caused diseases. Rhizoctonia on ornamental bedding plants can be controlled with soil drenches of Chipco 26019 50W (iprodione), or Terraclor 75W (pcnb).

Marketing of Bedding Plants

Marketing bedding plants requires time for researching potential customers, researching different products, selecting target markets and developing marketing strategies for gaining exposure and acceptance by consumers.

Management ability is a significant factor in the success of a bedding plant enterprise. The key management requirements are:

— the ability to research consumer tastes and preferences, research products and services and determine target markets in which customer needs can be met with the products available

— the ability to juggle several activities at once
— the ability to manage labor
— the ability to sell products and please customers
— the ability to control the financial affairs of the business and production operations

The market for bedding plants is expected to increase in size due to increasing disposable incomes, greater public knowledge about gardening and landscaping, and increasing awareness of the therapeutic value of gardening.

The focus of a bedding plant enterprise is to produce a salable product. This means producers must provide consumers with the right product at the right time and place.

Both established producers and new producers must be prepared to research consumer tastes and preferences for bedding plants, the different markets for their products and the various products and services that can be offered to customers. This market research must be performed before any production activities are started.

Bedding plant growers may be required to break their potential customers into smaller more easily identified groups. This practice of market segmentation is necessary when customer needs in a market become so diverse that choices must be made as to which customers will be catered to.

Markets can be broken into smaller more manageable groups based on geographical boundaries, demographic factors or product use factors.

Producers marketing bedding plants directly to consumers should be familiar with the different reasons (product use factors) customers have for buying bedding plants. A study of American gardeners suggests four distinct types of gardeners:

— Dabblers may be the largest market segment. These gardeners view gardening as a hobby and spend an average of two hours per week in their gardens and spend a relatively small amount on plants.

— Decorators spend an average of five hours per week in their gardens. This market segment view gardening as a means of improving the look of their homes.

— Cultivators spend an average of 10 hours per week in their gardens. The focus of their gardening activities is to produce fruit and vegetables.

— Masters are highly experienced gardeners who offer advice to others.

Bedding plant producers must be prepared to research all the possible products and services they might offer to customers in order to narrow their product mix to a manageable size.

By identifying market segments as well as possible products and services, bedding plant growers are able to determine target markets in which they are able to meet customer needs with their products and services.

Marketing bedding plants directly to consumers can occur through retail outlets that are part of the greenhouse facility or through a farmers market.

Marketing bedding plants directly to consumers through retail outlets located at the greenhouse facility allows growers to gain the following advantages:

— growers supply high quality products to their customers

— direct sales are cash sales with immediate payment

— growers interact with their customers and gain valuable feedback

— growers have some control over their prices

— growers are able to market from their property

Having a retail outlet at the growing facility allows producers to start small and test the market as they grow in size. Producers will also have the flexibility of being able to perform other tasks while the outlet is open for business.

Developing and operating a retail outlet at the greenhouse facility means the operator must deal with the following issues:

— There are a wide range of regulations that must be followed when operating a retail outlet for bedding plants. Regulations to be investigated include the following:

  — zoning and building ordinances governed by the local municipality

  — health regulations set by both provincial and local authorities

  — business license requirements set by local authorities

  — provincial or municipal sign laws that limit the size, location or type of sign used to promote a retail outlet

— Liability coverage is highly recommended to provide coverage against liability resulting from injuries to customers.

— Retail sales outlets will require proximity to large population centers, good roads, a parking area, good signage and facilities to accommodate customers.

Growers and marketers of bedding plants will need to develop merchandising strategies focusing on product selection, packaging, pricing, display and customer service.

Producers who choose to operate a retail outlet for their bedding plants must be prepared to perform the following activities and services that will increase the value of their plants to consumers and thus increase sales levels:

— Determine the promotional activities that will reach potential customers and generate awareness and purchases.

— Develop a retail environment that generates excitement and sales appeal.

— Present high quality bedding plants that are desired by consumers.

— Effectively display plants to enhance image and promote sales. An important element of displaying bedding plants is to provide ideas to customer for the use of the bedding plants.

— Use bedding plant containers that meet consumer preferences and retail needs rather than just production needs.

— Labels that identify plants by their names are essential information required by consumers.10 Labels that are pleasing to customers will increase the value of the plant to the customer.

— Services such as product information, wide selection and convenient hours will contribute to increased sales.

Marketing directly to consumers requires growers to be friendly and outgoing, have retailing skills, have the necessary product knowledge and be prepared to work long hours, seven days a week.

Growers who market bedding plants directly to consumers will be required to price their plants. Pricing is a critical element in achieving sales and achieving profitability for the business. The important issues in pricing bedding plants are:

— Cost of production is the minimum acceptable price for bedding plants will cover all costs including production costs, marketing costs, overhead costs, the cost of unsold plants, the cost of unpaid accounts and a return to the owners capital and management contributions.

— Competitor prices are an important consideration in determining where market prices are. Pricing at the same level as a competitor can be dangerous if costs of production aren't taken into consideration.

— Customer perceptions are also an important factor in pricing. Customer questionnaires should be used to determine how customers rank the importance of price compared to factors such as service and selection. An understanding of customer perceptions of value will allow retailers to determine a maximum acceptable price consumers are prepared to pay for their plants.

— Customer service is an important factor in increasing customer perceptions of value and therefore increasing sales levels. Customer suggestions for improved customer service include:

  — checking to see what people need or want

  — guaranteeing and replacing plants

— more advertising
— not misrepresenting products or services
— being staffed with well-trained people
— not practicing high pressure sales

Farmer's Markets

Farmer's Markets consist of a number of growers selling their produce directly to consumers at a common location. Each grower has a separate stall or stand at the market. There are farmer's markets located in most urban centers.

The advantages of farmer's markets are:

— Individual growers benefit from collective advertising that attracts more people to the market location.

— Farmer's markets provide new growers the opportunity to gain exposure with consumers. As well, farmer's markets provide a means for growers to market surplus produce that was not sold through their retail sales outlet.

— The market provides parking and good access to consumers that may not be available to individual growers. There are numerous markets operating in the growing season so producers have the opportunity to market their produce at a number of locations and increase their exposure to customers.

The disadvantages of farmer's markets are:

— the producer must harvest, package and transport their product to the market

— since there will be many producers offering similar products, individual growers must be prepared for intensive price competition for sales

— growers must be prepared to have unsold plants that will need to be returned to the greenhouse facility

Some of the pointers for marketing at farmer's markets are:

— keeping the same stall location helps establish a market presence

— sell only high quality products

— work to establish a reputation as a grower of good plants

Producers may attempt to increase their revenue and profit levels through increasing the scale of the operation, achieving cost efficiencies and marketing larger volumes of high quality, consistent plants to wholesalers and retailers.

Producers seeking to market bedding plants to wholesalers and retailers must understand the marketing strategies of different retail clients. For example, one retailer may peruse the strategy of lowest prices while another may peruse the strategy of only selling high quality plants.

Producers who sell to wholesalers and retailers must produce plants that have high durability and shelf life. Producers must also have production processes that allow the operator to produce the plants in containers specified by the retailer.

Producers seeking to market to wholesalers and retailers must be prepared to perform the following activities:

— make business calls with wholesalers, retailers and garden centres in order to develop markets for the products

— maintain contact with buyers and developing contacts with prospective buyers

— assess new trends in order to respond to changing consumer tastes and preferences

— sort, handle and package plants as required by each retailer or wholesaler

— be prepared to deal with several different markets each requiring specific products.

The critical marketing issuefor the manager of a bedding plant enterprise is to research the various markets in order to determine which crop(s) to grow, to identify a target market for the plants and to determine how the plants are going to be marketed. As well, growers must be prepared to stay aware of trends, consumer tastes and preferences. Furthermore, the bedding plant operator must be prepared to continuously work to provide the precise products required by their customers.

4

Culture and Maintenance of Ornamental Plants

An ornamental plant is a plant that is grown for its ornamental qualities, rather than for its commercial or other value. They are typically grown in the flower garden, or as indoor plants. Most commonly they are grown for display of their flowers. Other common ornamental features include leaves, scent, fruit, stem and bark. In some cases, unusual features may be considered ornamental, such as the prominent and rather vicious thorns of Rosa sericea. In all cases, their purpose is the enjoyment of gardeners and visitors. Ornamental plants may also be used for landscaping, and for cut flowers.

Similarly trees may be called ornamental trees. This term is used when they are used as part of a garden setting, for instance for their flowers, their shapes or for other attractive characteristics. By comparison, trees used in larger landscape effects such as screening and shading, or in urban and roadside plantings, are called amenity trees. For plants to be considered as ornamental, they may require specific work and activity by a gardener. For instance, many plants cultivated for topiary and bonsai would only be considered as ornamental by virtue of the regular pruning carried out on them the gardener, and they may rapidly cease to be ornamental if the work was abandoned.

Ornamental plants and trees are distinguished from utilitarian and crop plants, such as those used for agriculture and vegetable crops, and for forestry or as fruit trees. This does not preclude any particular type of plant being grown both for ornamental qualities in the garden, and for utilitarian purposes in other settings. Thus lavender is typically grown as an ornamental plant in gardens, but may also be grown as a crop plant for the production of lavender oil.

Herbaceous Ornamentals

This section focuses on herbaceous perennials, annuals and bulbs. Most plants in these groups are prized for their showy blossoms. All three groups have been extensively hybridised by commercial growers to change the following characteristics:

- flower colour and size
- foliage colour, size and texture
- hardiness
- ability to self-clean
- bloom time
- increased petal number
- fragrance
- disease resistance
- drought tolerance
- shade and sun tolerance.

Perennials

Perennials have roots that live through the winter. They are able to regrow above-ground parts in spring.

Although some perennials can last almost indefinitely, others may last only a few years. You should remember that perennial does not mean perpetual. However, perennials usually can be propagated asexually for additional plants.

Perennials have a shorter bloom period than annuals. Some will bloom continuously if spent flowers are removed. Others will rebloom later in the same season. Gardeners may come upon the term "half hardy." Half hardy means "marginally hardy." In severe winters these plants will not survive. But in mild winters with ideal conditions and sufficient mulch, half-hardy plants will survive.

Annuals

Annuals complete their growth cycle in one year. To propagate themselves from year to year, annuals produce great quantities of seeds. If the flowers of annuals are cut before seeds can form, more blooms rapidly grow to assure a seed crop. Annuals usually produce more flowers than other types of flowering plants. They also flower earlier in their life cycle. Many annuals produce blooms from planting to frost. Some annuals are perennial in warmer hardiness zones. Examples of perennials grown as annuals are geraniums, dahlias, snapdragons, mealycup sage, pansies and Iceland poppies.

Bulbs

All bulb-type plants have underground storage parts. Bulb-type plants can survive dormancy without nutrients or moisture. In this sense they are perennials. Some bulbs are planted once and last for years. Others last only a year or two. Nonhardy bulbs must be planted each spring, dug up in the fall and stored where they will not freeze.

Biennials

Biennials are another group of plants that are common in the home landscape. Many gardeners are familiar with hollyhocks, Canterbury bells, most foxgloves, snow-in-summers, and money plants. Biennials produce only foliage and roots the first year. The roots survive the first winter. During the second growing season, the biennial blooms, produces seeds and then dies. Biennials may reseed and appear to be perennial.

Gardeners usually purchase biennials from nurseries, rather than grow them from seed. This avoids wasted space taken by biennials during their first, nonflowering year of growth. Biennials require the same basic type of care as annuals and perennials.

Perennials, annuals and bulbs give the home landscape colour and foliage interest throughout the growing season. The broad range of bloom time and variety of flowers and foliage can produce continuous colour and interest in a well-thought-out plan. Dried flowers, fruit, seed heads and leaves may be left on plants for winter interest and as a source of food and shelter for birds.

Garden Design

The most successful gardens usually begin with a carefully planned design. If you forego planning, the results are often a haphazard collection of plants. Too many variations in shape, colour, size and texture create confusion. A drawing on paper is helpful for location of dormant bulbs as well as perennials.

Cultural Requirements

Cultural requirements of site, site preparation, soil, drainage, selection, growing from seed vs. transplants, and care are generally the same for perennials, annuals and bulbs. Any special requirements will be addressed under each grouping.

Site

Because there is a perennial, annual or bulb for almost every type of site, study existing conditions and choose the right plant for the site. Some plants may not bloom or may even die if the site does not match their cultural requirements. Use reliable reference books to determine the cultural requirements of each species. Knowing cultural and site requirements also can prevent unnecessary use of chemicals on plants misdiagnosed as "sick."

Sun

The following definitions explain what is meant by SUNNY, PART SHADE, and SHADE LOCATIONS.

— *Sun*: at least 6 hours or more of full sunlight

— *Part shade*: 1/2 day of sun, such as morning sun and afternoon shade. It also refers to lightly filtered sun through high tree branches.

— *Shade*: north-facing beds shaded by structures; woodlands with minimal direct sun exposure

There are perennials, annuals and bulbs that will grow in each of these situations.

Soil and Drainage

Most perennials, annuals and bulbs grow best in a loose, well-drained, moderately fertile loam with adequate moisture. Soils that drain poorly, especially in winter, can cause the crowns to rot, resulting in death of many otherwise hardy perennials. Poorly drained soils also will threaten the health of most bulbs.

The pH range for perennials is from 6.2 to 6.7, annuals from 6.3 to 6.7, and bulbs from 6.0 to 7.0.

Clay soils may need to be amended with organic matter to improve aeration and drainage. Add needed amendments according to soil test recommendations. A general rule for adding organic matter is to add 1 part organic matter to 3 parts native soil by volume. If drainage is poor, but the gardener wants to grow selections that require good drainage, consider raised beds. Species that grow in moist or wet sites are additional options.

Ideally, test the soil and prepare new beds in the fall for planting the following spring. However, new beds can be prepared in spring. It may take longer to obtain soil test results from the lab during the busy spring season. Allow at least two weeks for newly worked soil to settle before planting. If soil is not allowed to settle, crowns of new plants may become exposed, threatening plant health.

Perennial weed control is essential in a new perennial flower bed. Controlling weeds is easier when done prior to planting the bed. Work or till soil to a depth of 18 to 24 inches for new perennial plantings, 8 inches for annuals, and 12 inches for bulbs. Rock garden varieties or other shallow-rooted selections require only a few inches of prepared soil, but the site must be well drained.

Selection

Flower gardeners often have favorites. Favorite plants

that are not appropriate for the site will require site modification or an alternative selection. When selecting, consider plant shape, such as column-shaped varieties, "roundy-moundies" or horizontally creeping types. Consider plant height, spread, leaf texture, form and colour. Purchase plants or seeds from reputable nurseries that guarantee disease-free, named varieties. Larger sized, more mature plants will usually cost more.

Growing from seed vs. purchased plants

Perennials, annuals and bulbs can be purchased in several forms: seeds; dormant, bare-rooted plants; rooted plants in pots of various sizes; dormant bulbs and pot-grown bulbs. Many perennials can be grown from seed, but they are more often purchased as transplants. Annuals are even easier to grow from seed, but most gardeners choose to purchase transplants. Bulbs are usually sold in a dormant state.

Mail-order nurseries ship dormant plants, growing plants, and bulbs at the correct planting time in your area. If you cannot immediately plant new stock, bare-root plants can be heeled in to a protected section of the garden. They also can be potted up and grown in a cold frame.

Plant bare-root and potted plants in a hole large enough to allow spreading out of the root system. Soak the roots of bare-rooted plants in a bucket of water for 1/2 hour just before planting. Plant at the same depth as they originally grew. Look for the original soil line on the plant. When selecting transplants, choose those with dark green leaves, sturdy stems and side branches. Roots should be abundant, but not pot-bound. Plant immediately if possible. If transplants must be held, store in a shady area and water as needed.

Planting

Plants should be spaced a distance equal to their mature width. Ten-inch plants should be planted 10 inches apart. Clusters of 3 or more large plants or 5 or more small plants of the same colour usually make a better visual impact than those planted singly.

Dig a hole that allows transplants to be at the same soil level as in the container. Loosen and spread roots to encourage new outward root development. Water immediately after planting.

Fertilise based on soil test results. Use either a time-released fertiliser that will last all season or plan to apply water-soluble fertiliser two or three times during the growing season. Liquid 20-20-20 or 23-19-17 formulations should be applied biweekly to annuals. Follow product instructions for dilution rates.

Gardeners who grow their plants from seed will notice the wide difference in seed size. Some seeds are so small that they should be mixed with sand before sowing. This allows better control over spacing. Seed companies offer some very fine seed in pelletised form. Plants that have very small, dust-like seeds are often purchased as transplants because of the difficulty in handling these seeds.

Watering

Apply water when soil becomes dry. Newly planted transplants quickly wilt if conditions are dry and windy. Water young and newly planted plants more frequently than mature, established plants. Keep soil moist to a depth of 4 to 5 inches. If rainfall does not provide this moisture, apply water at a rate of 1 inch per week. Try to keep water off foliage. A 2-inch layer of mulch that does not touch the crown of the plant will help conserve moisture. A new layer of mulch may be applied each spring after the soil warms to maintain a 2-inch depth of mulch.

Special Requirement of Perennials

There are many reasons gardeners choose to grow perennials. Perennials do not need to be planted each year. An all-perennial garden changes throughout the season. Perennials are easy to divide and share with other enthusiasts.

Design challenges

Successful colour combinations and continual colour throughout the season are challenges facing gardeners who choose to grow perennials. Plot the blooming periods of the plants you choose on a linear calendar. This aides in developing continuous bloom and combining colour combinations successfully.

Large plants may be planted singly. Plant smaller plants in clumps or "drifts" of 3, 5 or 7. Avoid planting in precise rows. Informal shapes are easier to maintain than formal groupings. Curved lines, varied plant heights, masses and textures result in a more natural appearance.

Careful attention to foliage is important with perennials because few perennials bloom for more than a few weeks. Texture, form, height and colour should be combined carefully. In addition to a linear calendar, plotting the garden to scale will ensure that these factors will be combined successfully.

Growing from seed

Many perennials grown from seed take one or two years before they flower. If you have the patience, equipment and the space, grow perennials from seed; it is substantially less expensive. Many perennials may be seeded in June, then transplanted in early autumn.

Buying transplants

When purchasing potted perennials at a nursery, avoid selecting plants in bloom unless you must be assured of a specific colour. Remove any flowers at planting time to promote root establishment. The price of plants relates to the size and age of the plants. Newer cultivars are more expensive than cultivars that have been marketed for several years.

Transplanting

Gardeners, who enjoy collecting perennials, find that as plants mature they may wish to relocate some of them. Transplant perennials in early spring just after growth starts or in early September. To relocate perennials during the main growing season, remove all flowers and cut back foliage. Trimming of flowers and leaves reduces transplant shock. Water as needed for good establishment.

Fertilising

Perennials often stay in the same location for years. This means that selection and preparation of site for new plantings are vitally important. Organic matter worked into the soil at planting time can provide nutrients and improve aeration and drainage.

For perennials that require annual fertilisation, apply in spring as new growth begins. Broadcast 1 pound of 5-10-5 or 6-12-6 over 100 square feet of area. Make another 1-pound application 8 weeks later. Wash off any fertiliser that remains on foliage. Astilbe, chrysanthemum, delphinium, lupines and summer phlox are perennials that should be fertilised annually. It is always best to follow recommendations of a soil test. Test soil every 3 years.

Mulching

Winter mulching helps to prevent winter injury from HEAVING. After soil has frozen, apply a layer of coarse mulch. Four inches of pine boughs, pine needles or straw are suitable. Pull back mulch from plant crown as soon as new growth begins in spring. Winter mulching of perennials is especially important in the following situations:

- — when marginally hardy perennials are grown;
- — for newly transplanted or divided perennials moved in fall.

Supporting

Perennials with heavy flower heads or long thin stems may require staking. The weight of water or wind can easily damage flowers and stems. Examples of perennials that usually need stakes or supports include delphinium, foxglove, Achillea 'Gold Plate,' selected asters and mums. Stake when plants are about 6 inches tall so supports are hidden by foliage later in the growing season.

Deadheading

The removal of spent blossoms is called "deadheading." Many perennials will continue to bloom if faded flower heads are removed. This is especially true of dianthus, coreopsis and shasta daisy. Some perennials will bloom again later in the season. Examples of perennials that will produce a second flush of blooms if flower stalks are cut back to the ground include delphinium and false indigo.

There are different methods of deadheading. Some plants will rebloom on the same flower stalk. In this case, it is necessary to remove only the dead blossom. Campanulas and balloon flowers will rebloom if faded blossoms are removed. Other perennials are deadheaded by cutting off old flower stalks down to where the new bud shoots appear. Obedient plant should be deadheaded in this manner. When deadheading, look for the location of the second bud or set of buds.

Although faded flower heads should be removed on all plants, only some will produce new blooms. Others will not rebloom. Removing spent blooms diverts energy that would otherwise go into seed production.

Garden markers and record keeping are very helpful with certain perennials. It is easy to forget about perennials that appear late in spring or die back before the rest of the garden. They are often "rediscovered" by accidentally digging into them. Good candidates for markers are butterfly flowers, Oriental poppies and balloon flowers.

Dividing

Most perennials decline with age and should be divided every 3 or 4 years. New growth can be noticed on the outer edges of clumps. Divide perennials when you notice the following:

- — a dead center formed in the crown area with a ring of growth around it;
- — blooms becoming sparse and smaller;
- — plants looking obviously crowded.

Do not divide perennials when they are in bloom. Divide midsummer and late-summer bloomers in spring (April or early May) when growth is 2 to 3 inches high. Divide spring-blooming perennials in late August or September. To divide perennials, dig the entire plant and separate it into sections, or take divisions from the edge of the clump while it is still in the ground. Ornamental grasses have tightly tangled, tough root systems that are difficult to divide. Do not hesitate to cut through a large clump of roots with a sharp ax.

Some perennials are rarely divided or moved unless absolutely necessary. These include gas plants, peonies, tree peonies, foxtail lilies, goatsbeards, butterfly weeds and bleeding hearts. After a killing frost, remove dead stems and foliage. Leave a piece of stem a few inches long to assist you in locating plants the following spring.

Diseases and pests

Powdery mildew is probably the most troublesome disease of perennials. It looks like the plant is covered with a dusting of white or gray. It is very unsightly, even though plants are generally not killed by the infection. Powdery mildew usually appears in summer when fungal growth is promoted by high humidity.

To discourage powdery mildew, choose sites that are sunny and have good air circulation. Maintain good air circulation by thinning clumps and providing room for plants to grow. Water early in the day and remove and destroy any infected foliage. Powdery mildew can overwinter on infected foliage and spread to new foliage the following year.

Insects that can attack perennials include aphids, plant bugs, leafhoppers, mealybugs, scales, thrips and whiteflies, red spiders and other mites, caterpillars, grasshoppers, beetles, and leafminers. Slugs are often a problem, especially in shaded gardens where mulch is routinely applied.

Special Requirement of Annuals

Planting annuals runs a close second to lawn care as the most common home gardening activity. Abundance of flowers and a long blooming period, along with low cost, range of size, shape, colour, and adaptability make annuals garden favorites. Annuals listed in order of popularity with gardeners are geraniums, impatiens, petunias, marigolds, begonias, pansies, salvia and vinca.

Selection

Gardeners usually choose annuals for bedding or border plantings. They enjoy the mass of bright colours most annuals produce from spring to frost. Annuals also make excellent quick fillers where permanent plants have not yet grown large enough to give a mature appearance. Since most annuals have small root systems, they are easy to transplant and can be planted over spring-flowering bulbs.

Annuals for containers

Annuals grow well in containers, planters, and flower boxes. They can tumble down walls and soften sharp edges. Selections for these uses are described as cascading. Their growing habit causes them to trail over the edge of their growing space.

Some annuals produce fast, upward growth. Fast-growing annual vines can produce privacy screening or cover unsightly walls in just weeks. Annuals suitable for screening or covering include cathedral bells, hyacinth bean, moonflower, morning glory, scarlet runner bean and vining nasturtium. Some annuals are grown for drying and used in arrangements and wreaths. They are called "everlastings." Popular everlasting annuals include celosias, amaranthus, strawflowers, Gomphrena globosa and zinnias.

Plnating from seed

Some annuals must be started indoors during late winter or they will not have enough time to reach maturity. Annuals that have a long growing period before they reach maturity include lisianthus and ageratum. On the other hand, annuals that can be easily sown directly outdoors include marigolds, nasturtiums, cosmos, cleomes, California poppies, larkspurs, portulacas and sweet alyssums.

When growing annuals from seed, PINCH young seedlings to increase the number of blooms. When young plants are 2 to 4 inches tall and have 3 to 4 sets of leaves, pinch out the central stem as shown in the following illustration. This encourages side branches that produce a bushier plant that can support more flowers.

Grwoing from transplants

Annuals are most often purchased as plants in spring and planted after the last frost. Some annuals can take light frost and may be planted a little earlier. They include pansies, snapdragons, stocks and calendulas. Once annuals are properly transplanted into an appropriate garden site, they require minimal care. Remove spent blooms for continuous blooming. Some annuals are called self-cleaning. This means that they drop spent flowers, looking neat with less care. The impatien is an example of a self-cleaning annual.

Diseases and insect pests

Annuals occasionally become infected with the following diseases: leaf spot, powdery mildew, rusts and wilts. Common insect pests that affect annuals include aphids, spider mites, plant bugs, Japanese beetles and slugs. Refer to reference materials for plant-specific disease and insect problems.

Woody Ornamentals

Woody ornamentals are trees, shrubs and woody vines. They are the backbone of the landscape. Used creatively, they define outdoor living spaces. Trees and vines add a vertical dimension to the landscape. Woody ornamentals are more permanent than the herbaceous, nonwoody ornamentals. They usually cost more and require more work to plant properly. For these reasons, it is critical that a gardener plan before planting. Choosing the right plant for the right place will save much effort, expense and provide long-term satisfaction.

Planning

Planning requires knowledge of both the site conditions and the cultural requirements of plants. There are hundreds of species of trees, shrubs and vines available to the gardener. A gardener's knowledge grows with exposure to new plants and situations. Observing where plants do well in nature is helpful in expanding your basic knowledge of woody plants.

The first question to ask when thinking about planting a woody plant is, "Why is that plant going into that spot?" The plant must have a function in the location where it is to be planted. That function may be to look pretty. Looking pretty also could be stated as being an attractive specimen plant that draws the eye to a particular area of the landscape. A magnolia in full bloom can be used to lead the eye of a visitor to your front entrance.

Trees, shrubs and vines also are planted for practical functions. They are excellent for screening an ugly view. Vines can be planted and maintained in very narrow spaces and offer excellent screening. Trees and vines are often planted to shade a patio or the house. Well-placed trees and shrubs divide the landscape into areas for play, entertainment, specialty gardening and yard maintenance.

Environmental Factors

Hardiness zone, soil conditions, moisture level and exposure are key factors in site analysis. Knowing these key factors permits gardeners to select the right plant for each site. Problems arise when these factors are not considered. Decisions should be based not only on the plants a gardener likes, but also on the plants that are appropriate and will do well under the existing conditions.

Hardiness

Hardiness is normally based on the average low temperature of winter. This is a critical factor. If a plant cannot survive the coldest winter temperatures for extended periods, it is not reasonable to plant it. While winter protection can modify cold hardiness to some extent, it involves additional maintenance and the risk of the plant's death.

Winter or cold hardiness is not the only concern. Plants must also be able to survive other seasonal conditions: summer heat, drought, rainfall and drying winds. These and other weather conditions affect hardiness. A plant cannot be considered hardy for an area if it cannot survive during all seasons. However, hardiness maps are based on winter conditions.

Soil type

Soil affects plant hardiness. For example, a Zone 6 plant may be hardy in Zone 5 if it is planted in a sandy loam. However, if the plant is in heavy clay, that may retain too much moisture during winter, the plant's potential hardiness is reduced. Most references provide the recommended soil type for a particular plant. You will most commonly see "moderately fertile, well-drained loam" as the recommended soil condition.

Soil fertility and pH also affect plant selection and growth rate. Some plants require a regular application of fertiliser, while others survive quite well under low fertility. Likewise, particular trees and shrubs grow better in certain pH ranges. Rhododendrons, azaleas and fothergillas grow best in acidic (pH 5.5 to 6.2) soil.

Moisture

Moisture availability also must be considered. Too much or too little moisture may be a limiting factor when choosing a plant. Few plants can tolerate constantly wet soil. When soil is saturated with water, air is forced out. Plants cannot move oxygen from the leaves to the roots. Oxygen must be absorbed through the roots. However, some plants are adapted to wetter soil conditions.

Irrigation is an option for some dry sites. However, it can be expensive and labor-intensive and must be applied on a regular basis. Planning a landscape with differing zones of watering needs is a practical approach. Place plants with similar moisture needs together to simplify irrigation and avoid incompatibility problems.

Exposure

Exposure to sun and wind are important considerations in plant selection. Drying winds and reflected heat from buildings cause desiccation, especially during winter when the ground is frozen and moisture is unavailable to plants. Prolonged desiccation can result in death of individual cells and plant tissue. The leaves of broadleaf evergreens will curl to reduce the amount of leaf surface exposed to drying wind. In summer, drought combined with high temperature and drying wind will cause noninfectious or environmental leaf scorch on many plants. Leaf scorch can be intensified for trees that have roots growing under pavement. In this situation, moisture is blocked by the pavement. As the pavement heats, it raises the temperature of surrounding soil, causing it to lose more moisture.

The amount and intensity of sunlight influence plant growth and selection. Plants are often listed by sun and shade requirements. Placing a plant that requires a sunny exposure in a shaded environment may result in slow, leggy growth, reduced flowering and poor fall colour. Conversely, plants that naturally grow in shade placed in a sunny exposure are likely to show symptoms of desiccation and leaf scorch. There are various degrees of sun or shade. Full sun means 6 to 8 hours of exposure to sunlight. Filtered shade is sun coming through a canopy of leaves. Partial shade refers to an area that is shaded for part of the day and sunny for less than 6 hours a day. Deep shade is cast from a structure resulting in little or no direct sunlight.

Aesthetics

Aesthetics is the subjective evaluation of beauty. Plants have individual and collective aesthetic value. Some plants are so beautiful, they can stand alone as a focal point. Others are individually less beautiful, but can be combined with complementing plants to create a pleasing effect. Most landscapes are limited in the number of plants that can fit into the available space. Know the size of a plant at maturity and provide adequate space for it.

It is important to consider the aesthetic value of the plant throughout the seasons of the year. Too often, gardeners focus on the season of flowering and forget about the other seasonal qualities of a plant. Too many popular, spring-flowering plants have little else to qualify them for landscape use. Form, interesting bark, fall foliage and fruit are other important characteristics that can enhance the landscape throughout the year.

A plant that combines many of these characteristics throughout the year is called a four-season plant. Some flowering crabapples fall into this category. In spring, the flush of colour from the blossoms is spectacular. The summer foliage is glossy and green. During autumn the small apples form, providing a second season of colour. The best crabapple cultivars hold their fruit into winter adding aesthetic interest as well as food for birds.

Plant selection

With so many plants and so many design considerations, how does a gardener determine which plants to grow? You may choose to rely on the judgement of professionals. The alert gardener observes what is growing in the neighborhood and evaluates the positive and negative aspects of these plants.

Special forms

The shape and structure of the plant should blend well with the other plants and serve a purpose. Plants can be classified as rounded, oval, pyramidal, vase-shaped, upright, horizontal or columnar. Weeping trees, contorted shrubs and distinctive shapes will stand out from most landscape plants. These special forms should be used as focal points. Use contrasting form for added interest and depth. Too much contrast in form will result in a lack of harmony in your landscape. Plan for a plant's growth and mature form. Some plants change their form as they mature. The Scotch pine begins its life with a conical form, but matures into a windswept, irregular shape when mature.

Texture

Texture is frequently associated with foliage. However, the woody structure of a tree or shrub is also part of the texture. Fine texture is associated with small leaves and thin branches. Cotoneasters and willows are examples of finely textured plants. Coarse texture is achieved with larger leaves and thicker, sparse branches, as on oakleaf hydrangea.

Coarse textures are more appropriate for larger landscapes, while fine textures are more appropriate in a smaller landscape. A small area can be made to look larger by using finely textured plants. Coarsely textured plants should be used as accents in the smaller landscape.

Size

Landscapes should have plants of different sizes for added interest. Taller plants normally form a backdrop for progressively shorter plants. The scale of the landscape and permanent structures must be considered when choosing plants. A towering tulip tree dwarfs a one-story home on a small lot. The same tree is in scale with a three-story mansion on a large estate. When planning a landscape, allow enough space for each plant to reach its maximum size. A mature Scotch pine tree has a spread of 30 to 40 feet and reaches a height of 30 to 60 feet. The small holiday tree that is only 5 feet tall and 4 feet wide should not be planted 8 feet from the corner of the house.

Many people crowd young plants together to make a new planting look good when it is planted. This is fine if some of the plants are removed as the most desirable plants grow. Alternatively, annuals or less expensive, short-lived plants can fill the gaps until the woody plants begin to fill the space.

Growth rate

Slow-growing plants are ideal for rock gardens or for a collection of dwarf plants. Slower growing plants require less maintenance. However, most people want woody ornamentals to grow as fast as possible. There are drawbacks to fast-growing trees and shrubs:

- — They require more maintenance.
- — Their wood is weaker than slower growing species.
- — They are more likely to be damaged in high winds or during ice storms.
- — They often have a shorter life span.

Planitng Woody Ornamentals

Improper planting is a major reason for tree and shrub problems. A correctly planted tree or shrub that is appropriate for the site conditions usually will be a low-maintenance asset in the landscape.

Seasons to plant

Spring and late summer/early fall are the best times to plant trees and shrubs. Some plants are known to be difficult to transplant and these should be planted in the spring. Where conditions are unfavorable or difficult, spring planting is recommended. Spring planting provides a longer season for establishment.

Woody plants recommended for spring planting include birches, clematis, dogwoods, European hornbeams, hawthorns, goldenrain trees, magnolias, oaks, flowering pears, poplars, sourwoods, sweetgums, tulip trees, willows, Zelkovas, raspberries, blackberries, cherries, apricots and peaches.

Fall planting for the majority of trees and shrubs is excellent. The plant naturally requires less moisture at that time of the year. Shoot growth is minimal and is becoming dormant. Soil temperature is ideal for root establishment and usually remains high enough for root growth into December. Stored carbohydrates will be at a maximum in fall and available for root growth.

With fall planting, mulch the soil with 2 to 3 inches of aged organic matter to keep the soil warm for a longer time.

The planting hole

The planting hole will differ for bare-root plants. The hole should be dug so the plant will sit on a cone of undisturbed soil. The cone should support the plant so the flare at the base of the trunk is just above the soil line. Dig a hole that is 12 inches deep and extends out twice the length of the roots.

Balled and burlapped plants and container-grown plants should sit on a mound of undisturbed soil. It is critical that the ball be at the same depth as it was in the nursery. The mound for container-grown plants will not be at a depth for the container because you will be butterflying. The mound must be high to maintain the correct depth. The width of the hole should be at least twice that of the soil ball. The wider the hole, the better the root establishment.

The most common mistake is planting too deeply. The flare at the base of the trunk should be just above the soil level.

Backfill

Backfill is the soil that is dug out of the hole and then used to place around the roots when planting a tree or shrub. Most soil should be amended before it is used as backfill.

The soil, whether clay or sand, usually lacks organic matter. The soil should be improved by adding up to 1/4 by volume of organic matter to the native soil. The organic matter can be humus, aged manure, peat moss or a combination. Do not use fresh manure, grass clippings or uncomposted material.

Fertiliser

A soil test should be done before planting so that lime and phosphorus can be added to the backfill. Other nutrients may be added after several months when the plant has become established.

Watering

Fill the hole about halfway with backfill and then irrigate to settle the soil. Then add backfill to ground level and water again to settle the additional backfill. Additional backfill may be required once the soil settles. This process eliminates air pockets. Do not pack wet soil with your feet or a tool. Allow it to settle naturally.

Watering will be made easier if soil is piled around the planting hole to create a saucer to hold water. During the first year after transplanting, irrigation is important. Apply enough water to thoroughly wet the root ball. Check by probing the soil to make sure the soil ball is getting wet. The amount of water applied will depend on the amount of rainfall. Apply water slowly with a dripping hose or a soaker hose to avoid runoff.

Pruning

Only minimal pruning should be necessary when planting. Remove broken branches and branches that are rubbing. Bare-root plants may require additional pruning because of the reduced root surface. One year after planting a tree or shrub, prune for proper structure and shape.

Tree wrap and support

Trees with thin bark (dogwoods, maples, birch and fruit trees) that may be damaged by sun or frost may be wrapped. Use tree wrap paper and wrap from the bottom up, overlapping slightly. Remove loose wrap and plastic tubes that may be on new trees. Excessive moisture is maintained behind these resulting in fungal growth. Insects may become a problem under loose wrap. Remove all wrapping material after the first complete growing season. Wire mesh may be placed around the base of the trunk to prevent rodent feeding during the winter.

Staking is only recommended for bare-root trees and in problem situations where a tree is likely to be blown over. When staking is necessary, it must be done correctly and left in place no more than one year. Guy wires must provide some movement of the tree. Where guy wires wrap around the tree, they must be covered to avoid damaging the bark.

Maintance

A regular programme of maintenance is required for best results with any tree or shrub. This programme includes irrigation during the early years, core aeration, mulching, fertilisation, pruning and regular inspection for disease and insect problems.

Watering

Watering woody ornamentals depends on the species and the soil. Some plants tolerate dry conditions better than others. A tree growing in a fertile, well-drained sandy loam may become established in a year or two. This tree would require little supplemental irrigation. A shallow-rooted tree planted in a poorly drained, heavy clay may require supplemental irrigation for several years.

During the first 4 to 10 years after planting, trees and shrubs may require supplemental watering. After that, the amount of water required for a mature tree or large shrub is so excessive (40 to 50 gallons of water per day per tree) that supplemental watering may not be realistic.

The most critical season for supplemental irrigation is usually during summer. However, any time during the growing season when rainfall is inadequate for more than a week or two can create water stress. It is necessary to begin watering before soils dry completely. Probing the soil and feeling for moisture is a reasonably accurate way to determine soil moisture. Water stress in late fall is not normal and has dire results when it occurs. Plants must have adequate moisture to tolerate the stresses of winter.

Core aerating

The roots of trees and shrubs soon grow beyond the planting hole into native soil. In areas where building has occurred, topsoil is usually removed, leaving clay subsoil. Aeration and drainage are a problem. Grass roots compete with tree roots for nutrients, moisture and air. Deep core aeration lessens these problems.

In deep core aeration or vertical mulching, drill holes into the soil beginning 6 to 10 feet from the tree trunk and extending 4 to 6 feet beyond the longest branches. The holes are 1 to 4 inches in diameter, 12 to 18 inches deep and 1 to 4 feet apart. Removed the soil from the hole. One- to 2-inch diameter holes do not need to be filled. The wall of the hole will gradually collapse creating additional air space in surrounding soil.

The 3- to 4-inch diameter holes should be filled with composted organic matter. The organic matter can be mixed with an equal part of coarse aggregate, such as vermiculite.

Deep core aeration should be done when roots are actively growing. Early spring and late fall are good times to core aerate. The soil should not be excessively wet or dry. Core aeration is not necessary until the roots have grown out of the planting hole into the surrounding soil. As a regular maintenance activity, core aerate every third year. For trees under stress, more frequent aeration may be beneficial.

Mulching

Apply a mulch 2 to 3 inches deep over the soil. Keep the mulch at least 3 inches away from the base of the trunk. The mulch will help retain moisture, moderate temperature extremes and reduce the potential for physical damage to the base of the trunk.

Mulching trees and shrubs has several benefits. Mulches retain moisture, suppress weeds, moderate soil temperatures and reduce physical damage from lawn mowers and string trimmers. Organic mulches slowly decompose adding organic matter to the soil. Organic mulches should be periodically renewed. Never exceed a total depth of 3 inches of mulch.

Fertilising

Soil and environmental conditions in a landscape may differ from those in a natural setting. The soil in most landscapes has been disturbed. It is often subsoil from excavation of the foundation. In a forest, leaves fall to the ground each year, returning nutrients to the soil. Growth is much slower in a natural setting, resulting in lower nutrient needs. Trees and shrubs often must compete with turfgrass for nutrients.

Landscape trees and shrubs must be fertilised for optimal growth. Application should be in late October through early December or from February through May, while the plants are dormant. Frequency of application is dependent on soil conditions. Annual fertilisation is recommended in poor soil, while every other or every third year is adequate in good soil.

The fertiliser should have a ratio of 3-1-1 to 4-1-2. Availability of products vary. Look for 24-8-16, 18-6-12, 18-5-9, 15-5-5, 12-4-4 or a similar ratio. Application rates should be based on a soil test. However, a general rate is 1 pound of nitrogen per 100 square feet for shrubs. The rate for trees will vary depending on application method and competition from turfgrass. If turfgrass is growing under a tree, apply the fertiliser in core aeration holes or use liquid fertiliser injected into the soil. For trees that are mulched or have ground covers under them, see the following general recommendation:

— For trees under 6 inches DIAMETER BREAST HIGH (DBH) apply 1/4 pounds of nitrogen per every inch DBH.

— Trees over 6 inches DBH: 1/2 pound of nitrogen per inch DBH.

For instance, applying 1/2 pound nitrogen per inch DBH to a 6-inch DBH tree using an 18-5-9 analysis fertiliser would be calculated as follows:

— 6″ (diameter) X 0.5 (rate) = 3.0 (amount of nitrogen)

— 3.0 (amount of N) / 0.18 (%N in 18-5-9) = 16.7 pounds of 18-5-9 fertiliser.

The fertiliser should be applied evenly under the entire CANOPY of the tree. Applications early in the growing season can be split into 2 equal parts applied 6 weeks apart. Fertiliser spikes are popular alternatives. Carefully read the package label to approximate application rate and nutrient analysis.

Commercial tree care firms offer nutrient injections. These are recommended only when the conventional methods are not possible. Trees surrounded by pavement are an example of when this may be necessary.

Pruning

Pruning is the selective removal of plant parts: shoots, branches, roots, flower buds, seed pods or fruits. A regular schedule of maintenance pruning will reduce growth problems. Proper pruning will enhance a plant's appearance and health, while improper pruning will result in serious health and aesthetic problems.

Plant health is a primary reason to prune. Dead, diseased or injured parts should be removed. A dense canopy should be removed to improve sunlight penetration and air circulation. Crossing branches should be removed to prevent rubbing of the bark.

Training young plants will result in an excellent structure of branching. Branches should be well spaced and at the strongest angles. Creating unnatural forms through pruning has been practiced for years. BONSAI, TOPIARY, ESPALIER and POLLARDING are high-maintenance endeavors that can create dramatic focal points.

Controlling growth is often necessary. Pruning to control the size of a plant results in high maintenance. An alternative is planting a species that only grows to a specific size. Rejuvenating old shrubs is a challenging pruning project. A sparse or overgrown hedge can be rejuvenated by pruning back in late winter.

Improving appearance is a legitimate reason to prune. Suckers at the base of a shrub are ugly and should be removed. Avoiding injury or damage by pruning is wise maintenance. Low branches can damage cars, poke people in the eye or be a hazard for the person mowing the lawn. Limbs hanging over utility lines, the house or other valuable property can be a liability.

Flowering and fruiting can be controlled by pruning. Flower buds can be removed to encourage more vegetative growth. Pruning can be used to stimulate flowering on some plants like wisteria and lilac. Selectively removing some flowers will result in others being larger or producing larger fruit.

Pruning equipment

The basic tools every landscape gardener should own are hand pruners, lopping shears and a pruning saw.

Hand pruners or pruning shears are used to clip off stems up to 3/4 inch in diameter. Scissor types with two sharpened blades that overlap when making the cut are recommended. The anvil pruners have one sharpened blade that cuts against a flat plate. Anvil pruners tend to crush the bark of the stem being cut. Ratchet or geared pruners add leverage and should be used to reduce the muscular effort required to make a cut, not to cut larger stems.

Lopping shears have long handles to provide leverage to cut through branches up to 1-3/4 inch in diameter.

They are useful in rejuvenating shrubs. A narrow, curved pruning saw is excellent for removing limbs up to 2-1/2 inches in diameter. A pruning saw can fit into tight spots and make clean cuts. Larger cuts should be made with coarse-toothed saws or chain saws. If the effort to cut any stem is so much that the person resorts to twisting, the next larger tool is recommended. Follow safety precautions when using tools, because the injuries that can result from accidents are serious. Hedge clippers or shears are not used for making pruning cuts. They make straight-line cuts without regard to nodes.

Pruning deciduous trees

All pruning cuts should be made just above a node. Your cut may go back to a bud or to where a branch emanates from a larger branch. Leaving a stub results in the stub dying. These stubs are entry points for disease and insects.

When a cut is made, hormones are released that cause buds within a few inches of the cut to break and grow. This gives you great control over where to make a cut, since you can determine which buds are likely to sprout. Where possible, prune above an outward-facing bud.

The appropriate angle for a branch coming off the trunk is a 45- to 60-degree angle. This results in the greatest strength for the limb.

When sawing off limbs that are greater than 1 inch in diameter, begin by making an undercut about 2 inches out from the final cut. Then cut into the top of the branch slightly out from the first cut. Then make the final cut just out from the branch or ridge collar. This will prevent the branch from tearing bark from the trunk of the tree.

Begin pruning a tree by removing dead, diseased or broken branches. Then remove branches that go straight up (water sprouts) or point downward. Then look at the branch structure and remove branches that are or will eventually become crossing branches. The final pruning will be based on total tree structure and appearance.

Pruning deciduous shrubs

Multistemmed deciduous shrubs benefit from the removal of the oldest stems at the base of the plant. This keeps the plant open and vigorous. Follow this practice on lilacs, forsythias, and mock oranges. Allow a few of the most vigorous suckers to remain as replacements for the old stems. For shrubs that have not been properly pruned for many years, remove no more than 1/3 of the largest stems in any one year. Continue this procedure for 2 more years to complete the rejuvenation.

Other maintenance pruning includes the removal of dead, diseased, crossing and broken branches. Long branches can be cut back to a bud or to the point of attachment of an outward growing branch.

Prune most deciduous shrubs in early spring before bud break. Pruning in early spring will remove the flower buds of spring-flowering shrubs. Therefore, prune spring-flowering shrubs after the blossoms fade.

Pruning broadleaf evergreens

Broadleaf evergreens generally require very little pruning. Pruning should be limited to selective pruning to improve the natural habit or to keep it in scale with surrounding plants. Prune after the flowers fade. Prune back to a branch or bud. It is recommended to remove spent flowers of azaleas, mountain laurels and rhododendrons by snapping them off by hand.

Rejuvenate large or leggy, broadleaf evergreens in early spring. Cut back 1/3 of the largest branches within

6 inches to a foot of the soil. Continue this process over the next two years. Apply fertiliser after rejuvenation pruning and before buds begin to grow.

Pruning hedges

Training plants into hedges must begin when the plants are planted. Always keep the top of a hedge narrower than the base, so all foliage is exposed to sunlight. Hedges may be sheared for a formal look or pruned to maintain the natural form of the plant. The formal shearing must be done 1 to 3 times a year to maintain the formal appearance. The first shearing should be done shortly after spring growth begins. Then repeat the procedure in early July.

Pruning vines

The primary reason to prune vines is to limit vigorous growth. Most pruning should be done in late winter or early spring while the plant is still dormant. Vines that climb by HOLDFASTS should be pruned to remove dead wood or to limit growth. Twining vines are generally pruned to maintain a full covering of foliage. They tend to produce most of their foliage on the newer growth. Wisteria and clematis require special pruning techniques.

Transplanting

Moving established plants in your landscape is a physically difficult task. When a tree or shrub is dug, as much as 98% of the roots are left behind or damaged. Root pruning can be done a year in advance to form a compact root ball of fibrous roots. However, it is still difficult to dig a solid soil ball with a shovel. Begin by tying branches to prevent damage. Dig a trench around the area that is to be the soil ball. Undercut the soil ball. Slide a piece of burlap or heavy cloth under the soil ball and tie it around the trunk of the tree or shrub. Then slide the soil ball onto a snow saucer or similar device to make moving the plant easier.

Production of Ornamental Vegetables

Vegetables are often grown as a utility crop; planted in a less visible part of the garden, then harvested and used as a source of home-grown, tasty greens. Lately, there has been a development towards growing vegetables in a more ornamental way, opposed to the more common method of straight lines of one individual variety. Perhaps "development" is not the correct word to use because vegetables were grown in an ornamental and formal fashion as far back as Egyptian times. One of the oldest depictions of an Egyptian garden shows a grid-like pattern of small vegetable and fruit beds intermingling with flower beds and with a network of paths running between them. It is this method of vegetable cultivation to which there appears a natural regression.

It is a style that makes a lot of sense. The small beds can be tended without ever feeling the effects of soil compaction caused by the feet of well-intentioned gardeners, and weeding smaller areas is less of a psychological hardship than setting out to tackle one large, unwieldy expanse. At the height of the growing season, a group of small beds can look extremely attractive. These beds can be created to varying degrees of sophistication. The simplest approach is to have a group of small beds with sand, grass or gravel paths between. Raised beds are even more manageable and are particularly effective if edged with stone.

In the small beds, vegetables can still be grown as a crop to be harvested, or if the space is available, it is interesting to grow varieties with unusual colours or shapes, for example, rhubarb chard or bronze fennel. The miniature French bean, Purple Teepee, is eye-catching when grown closely together in a small bed, but remember that they lose their colour and revert to green when boiled. For an extra splash of colour, the yellow courgette, Gold Rush, provides brightness, and the large leaves are attractive as they spill over the edge of a small bed. More enhancement of a vegetable garden can be achieved by the addition of height - sweet-corn is excellent for this purpose. Wigwams with runner beans scrambling up them are also successful, especially if a bright flowering varieties of bean are chosen, for example, Scarlet Emperor and Red Rum.

Allowing small groups of vegetables to flower is also interesting. Red lettuce, Lollo Rosso, produces tall, pyramidal spires of small, pale pink and green leaves, and when radish is left undisturbed, it becomes very pleasing to the eye - a haze of white flowers on tall stems.

Vegetables are often attractive plants. Their mixture of leaf colours and textures add beauty to the staid vegetable garden. Ornamental vegetable gardening is not a new idea. Potagers, or decorative kitchen gardens, were a part of the elaborate gardens at Versailles during the reign of Louis XIV in the 1600's. The combination of aesthetics and practicality in an edible garden does not require the wealth of kings; only a little imagination and preplanning.

Potagers are often geometric in layout. They are also often edged with ornamental plants. Many European gardeners edge their potagers with boxwood. However, almost any small mounding plant will work as long as it does not hinder the growth and production of the vegetables within. The vegetables should receive full sun at all times or production can be compromised. Try Dalhberg daisies or marigolds for a blooming edge.

Potagers tend to be formal in design with pathways dissecting them. Trellises and archways are common features in many decorative kitchen gardens. They provide support for vining crops and add height to the garden. Vegetables and herbs with colourful foliage and varying textures are used to add more interest. 'Bright Lights' Swiss Chard, 'Purple Ruffles' Basil, 'Bronze' Fennel, and several varieties of leaf lettuce are just a few plants with colourful leaves or stems.

Ornamental Cabbage and Kale

Ornamental cabbage and kale have been the traditional ornamental vegetables for providing colourful and attractive foliage during the fall and winter months. Recently leafy vegetables, in particular oriental mustards, have become popular for landscape plantings during the cool season. Leafy vegetables which have been popular in salads and stir-fry dishes are now being adopted by curators of botanical gardens and landscape contractors as specimen plants, border plants, and in mass plantings. Planted with pansies and garden mums, these vegetables offer a change of texture and foliar colour. The shades of green and purple provided by these vegetables help to increase the colour in cool season gardens during October through March. Ornamental vegetables also produce brilliant yellow flowers in early spring that provide additional colour.

The increased interest in heirloom plants have added to the appeal of these vegetables including red mustard 'Red Giant', potherb mustard 'Early Mizuna' mustard-spinach type plants like flat pak choi or 'Tatsoi', and kales like 'Red Bor', 'Red Russian', and 'Lacinato'. Swiss Chard is another ornamental vegetable that provides colourful stems and robust foliage during the winter and warmer periods. Many greenhouse growers are now adding ornamental vegetables to their annual production cycle since they are easy to produce. The cultural requirements of mustards, kales, and Swiss chard are discussed together and specific requirements are covered in Table 1.

Selected Cultivars

Red mustard or India mustard

(Brassica juncea var. rugosa) produces large leaves with thick petioles. The cultivar 'Red Giant' has large, bronzed leaf blades that become more intense in colour as the plant matures. The leaf blades taper down to the main growing point that is thick and fleshy. The midribs are bright white and as the plant reaches full maturity, the leaves change from a rusty red to a reddish black colour. The colour of the leaves also changes from a dull to a shiny appearance. The application of plant growth regulators enhances the colour intensity. The plants rapid growth enables growers to market the plants 6 to 8 weeks after sowing. The roots are fibrous and the irrigation requirements increase as the root mass develops in the pot.

Red mustard can achieve a height of three feet as long as winters remain warm and wet. Temperatures below 15°F can damage the unprotected outer leaves and cause them to shrivel. The interior leaves of this biennial can survive low temperatures, but the entire plant may be reduced to only 5 to 6 leaves. To limit cold injury, the plants should be planted under trees or wind protected areas. The tall mounding growth habit makes 'Red Giant' an attractive specimen plant or background texture plant.

Potherb Mustard 'Early Mizuna' can be used as a specimen item in the landscape as well as a groundcover.

Table 1. Growing requirements for ornamental vegetables.

Crop	*Pot Sizes / Spacing*	*Fertility*	*Plant Growth Regulator Foliar Sprays*
Ornamental Mustards 'Red Giant' 'Early Mizuna'	4.5-inch pots on 10-inch centers 8-inch and gallon pots on 17-inch centers	200 ppm N constant feed	2,500 ppm B-Nine, multiple applications; 5,000 ppm B-Nine, single application
Ornamental Cabbage and Kale	4.5-inch pots on 10-inch centers 8-inch and gallon pots on 17-inch centers	200 ppm N constant feed	2,500 ppm B-Nine, multiple applications; 5,000 ppm B-Nine, single application; 2 to 8 ppm Sumagic, single application
Swiss Chard 'Bright Lights'	4.5-inch pots on 10-inch centers 8-inch and gallon pots on 17-inch centers	200 ppm N constant feed	
		200 ppm N constant feed	2,500 ppm B-Nine, multiple applications
Forage Kales 'Lacinato' 'Red Bor' 'Red Russian'	4.5-inch pots on 10-inch centers 8-inch and gallon pots on 17-inch centers	100 ppm N constant feed	2,500 ppm B-Nine, multiple applications; 5,000 ppm B-Nine, single application
Mustard-Spinach Type Vegetables 'Tatsoi'	4.5-inch pots on 10-inch centers 8-inch and gallon pots on 12-inch centers		None required

'Early Mizuna' has a mounded form and can reach 2 feet in the garden. The feathery-textured leaves are yellow-green and deeply notched. Plants may possess up to 180 leaves and all petioles are attached to a thick, tan stem. Plants may survive light frosts, but are killed by subfreezing weather. After a vernalisation period has occurred, the plants produce several elongated stalks, which contain the reproductive structures. 'Early Mizuna' is very attractive in bloom with a profusion of bright yellow flowers covering the plant.

Mustard-spinach type crops like pak-choi are also popular Asian vegetables. The variety 'Tatsoi' *(Brassica rapa var. rosularis)* produces firm, broad, fleshy dark-green leaves that are flat and shiny. The leaves have white petioles that attach to a central growing point forming a rosette. The low growing habit (up to 8" tall) of 'Tatsoi' makes the plants attractive in borders or in mass plantings. 'Tatsoi' does not survive temperatures below 25oF and bolts rapidly as temperatures begin to increase in the spring. 'Tatsoi', with its small compact form, can be grown tightly on benches or on ground beds like mums.

Forage kales *(Brassica oleracea acephala)* were originally were used as a source of fodder for cattle. 'Lacinato', 'Red Bor', and 'Red Russian' are three of the varieties used by botanical gardens today as ornamentals. 'Lacinato' grows to a height of 2 feet and has thick and crinkled dark blue-green foliage. 'Red Bor' achieves a height of 3 feet and possesses dark purple ruffled leaves. 'Red Russian' is an interesting variety that grows to 3 feet. As the cooler weather sets in, its foliage changes from a blue-green to a purple-red colour.

Swiss Chard *(Beta vulgaris)* is a relative of the beet, but unlike the beet, it is grown only for its colourful leaves and stalk. The plants generally grow 1 to 1 ½ feet tall. The cultivar 'Bright Lights' was a 1998 All-America Selections Winner. This variety is a favorite among gardeners because of its multiple coloured stems and robust and crumpled foliage.

Ornamental cabbage and kale *(Brassica oleracea L. var. acephala)* are also popular plants used in the fall landscape. Another name used to identify this crop is flowering cabbage and kale, but this label is inappropriate, because the plants are grown primarily for their colourful fall foliage and not the yellow flowers that appear in the spring.

Cultural Information

These plants, with their rapid growth rates, generally require 6 to 8 weeks of growth from seeding to achieve marketable size in 8″ mum pans or 1 gallon containers. Due to their rapid growth rate, ornamental vegetables can be of wholesale shipping size by 6 weeks, and retailers can market the crop at 8 weeks.

Germination and Plug Culture: Seed can be sown in plug flats, germination trays or into a 606 or an 804 flat. Seeds can also be sown in the final containers and then thinned when the seedlings begin to produce true leaves. Southern growers commonly purchase plugs (typically 288s or 384s) from northern plug producers. Germination takes 3 to 5 days at 70oF with the seeds covered lightly with germination substrate. If growers decide to sow these leafy vegetables themselves, the germination substrate needs to remain cool and moist. An application of B-Nine at 2,500 ppm two weeks after sowing can be applied to control plant height in the plug tray. The plants should not be allowed to become root bound in the germination container before transplanting, as restriction of the roots can result in stunted plant growth.

Seedlings should be transplanted into larger pots containing a substrate with excellent water holding capacity. Normally, growers will transplant the plugs into 6-, 7-, or 8-inch pots when the root ball is fully developed and the first true leaves are twice as large as the cotyledons. If plants become too tall in the seedling flats, they can be planted slightly deeper, up to the first set of leaves. These vegetables are heavy feeders and competition for nutrients should be avoided, so one seedling should be transplanted per finished pot. Swiss chard is the exception, avoid small celled containers and plant 3 to 4 plugs per 8-inch pot for a robust, colourful display.

Spacing: Due to their robust foliage, plants should be adequately spaced in order to maximise airflow to prevent foliar diseases and allow for uniform growth around the pot. If plants are being grown in greenhouses, extra care should be taken during watering, spraying, and shifting plants because the foliage may be fragile due to the lack of cool temperature acclimation. Use 6-inch centers for 4.5-inch pots, 11- to 12-inch centers for 6-inch pots, and 16- to 18-inch centers for 1-gallon containers and 8-inch mum pans.

Irrigation: Plants can be grown in a soilless substrate with good water holding capacity. If the plants are to be grown outdoors in pots, a soilbased substrate or the addition of sand may keep plants from being knocked over during windy weather. During the late summer months, when the plants are establishing themselves in pots, an adequate supply of water should be provided to the substrate. Use of an automated watering system similar to one used on garden chrysanthemums is recommended. Ornamental vegetables are not tolerant of water stress, which results in the stalling of growth, yellowing, and dropping of lower leaves. Premature bolting can occur if moisture levels become low in the growing substrate and extensive wilting takes place.

Fertilisation: Maintain the root substrate pH between 5.8 and 6.5. Irrigation water alkalinity levels should be below 2 meq/L of alkalinity (100 ppm $CaCO_3$ or 120 ppm HCO_3-) to avoid a general increase in pH over the growing season. If alkalinity levels are higher, consider using acid injection or an acidic fertiliser (as long as the NO_3-N to NH_4-N ratio is greater than 2:1).

Fertilise the plants with a balanced fertiliser of calcium nitrate ($Ca(NO_3)_2$) and potassium nitrate (KNO_3), with periodic applications of 20-10-20 or 21-5-20 (for phosphorus and micronutrients), and epsom salts [$MgSO_4 .7H_2O$] (for magnesium and sulfur). High levels of NH_4-N and/or urea in a fertiliser mix will stimulate stem elongation, resulting in the need for higher plant growth regulator rates to control plant growth. The percentage of ammoniacal nitrogen and urea should be less than 20% of the total nitrogen supply to avoid stretchy growth. For the seedling stage, fertilise at the rate of 50 to 100 ppm of N with a constant liquid feed. After transplanting into the final container, fertilise with a balanced fertiliser at the rate of 150 to 250 ppm N and K. Electrical conductivity (EC) should be maintained between 1.0 and 2.5 mS/cm (saturated paste extraction method) during periods of active growth. Excessive fertilisation will prevent good colouration, but deficient levels of fertility will result in yellowing (nitrogen deficiency), purpling (phosphorus deficiency), or defoliation of the basal leaves. Deficiency symptoms are more likely to occur in a soilless substrate, which has a low cation exchange capacity (the ability to hold nutrients), than in a soil-based substrate. Therefore, for the final two weeks of growth, a fertilisation rate of

100 ppm N and K should be sufficient while allowing the plants to develop colour and avoid deficiency symptoms. Temperature. Outdoor production of ornamental vegetables is highly recommended, as cool night temperatures in the late summer and early fall strengthen the foliage and may enhance foliage colour. When ornamental vegetables are grown in greenhouses, night temperatures should be held at 55 to 60°F during the final 4 weeks of production. Cooler temperatures will help avoid the leaves from becoming too soft and fragile and will allow the plant to harden and become acclimated to cooler outdoor temperatures. The exception is ornamental cabbage and kale, which requires night temperatures of 45 to 50°F for significant colour development.

Growth Regulators. Plant growth regulators (PGRs) can only be applied to ornamental vegetables grown as non-food crops. PGRs can be used to avoid stretching and achieve desired wholesale and retail plant heights and diameters. Based on PGR studies conducted at NCSU, multiple foliar applications of B-Nine at 2,500 to 5,000 ppm should be used to control plant height and diameter of 'Red Giant', 'Early Mizuna', the forage kales, and ornamental cabbage and kales. Multiple foliar applications of B-Nine at 2,500 ppm are appropriate for Swiss chard. Mustard-spinach is naturally compact and does not require PGRs. Lower PGR rates produce plants that are more robust and suitable for retail sales. Higher rates result in more compact plants suitable for wholesale shipping. Sumagic foliar sprays at 2 to 8 ppm were also found to be suitable for ornamental cabbage and kale in trials at NCSU, but Sumagic trials have not been conducted on the other vegetables. Besides controlling internode stretch and leaf expansion, PGRs also produce darker foliage colour on most ornamental vegetables and intensify the bronze appearance of 'Red Giant'.

Ornamental vegetables treated with PGRs should never be sold as edible crops.

The varying textures and foliar colours make ornamental vegetables an exciting, new addition to the fall garden. When the garden mums begin to fade and the pansies start their winter hibernation, mustards, kales, and chard can withstand the chilly air, and maintain colour throughout the winter months. The landscaping industry is always searching for attractive additions to their fall planting list. Growers should consider adding them to their fall production regime.

5

Flower Forcing Techniques

Cut flower production is a growing business in most countries as the result of higher per caput income of people. Technology for production is well known and has been utilised to the maximum in many countries. As a consequence, cut flowers are commercially produced in larger quantities in these countries resulting in surpluses during the normal periods of blooming, the end result of which is a lower price for cut flowers. On the other hand, such prices are much higher during the periods of needs such as certain special occasions, or off-season.

Flower forcing is any operation to induce flowering at a desired period with the goals of producing flowers during off-season and specified date, and for sale at higher prices than during normal blooming season. It has been hypothesised that, after reaching a ripeness-to-respond stage, and upon receiving proper stimuli, flowering hormones, known hypothetically as vernalin and florigen, are produced which stimulate the initiation of flower primordium and subsequent blooming.

The goals of flower forcing operation are as follows:

— *Off-season production*: Cut flowers that are available during their normal season are plentiful, thus fetching quite a low price. Sometimes, the farmers have to sell their flowers at a loss, otherwise none can be sold. In some cases, harvested cut flowers that could not be sold were left to spoil. Some were even left on the plant as it was not profitable to harvest them. This is a big loss to the farmers. Thus, it is advisable for the farmers to produce cut flowers during the off-season period, to avoid the loss and to obtain higher prices of the produce, although the inputs may be high.

— *Specific-date production*: Similarly, the demand for cut flowers is quite high during certain occasions, such as Christmas Celebration, New Year's Day, Mother's Day, Memorial Day, Valentine's Day, Graduation Exercise Day, Buddhist Lent's Day, etc. Thus, it will be to the farmers' advantage if they could produce cut flowers to be available on these occasions.

Plants behave differently with respects to their flowering time. Some flower all year round with either little or great seasonal influences, while others flower during their respective seasons. Flower forcing operations can be done through:

(i) adjusting factors which affect flowering,

(ii) the use of chemicals, and

(iii) mechanical forcing.

Flower forcing can be achieved by adjusting the factors effecting flowering behaviour, viz. photoperiod, temperature and humidity.

Chemical flower forcing can be done through the application of fertilisers, used both for retarding and stimulating flowering; and plant hormones, including gibberellin, growth retardants and growth inhibitors.

Mechanical flower forcing is achieved by operations such as, pruning, leaf trimming, ringing, budding or grafting, smoking, low-temperature storage and breaking dormancy.

The Flowering Process

Before a plant can flower in response to its environment (particularly temperature and day length) the organs that detect the environmental change, usually leaves or meristems, must reach a condition called 'ripeness to respond'. It has been hypothesised that, after reaching the "ripeness-to-respond stage", and upon exposure to proper stimuli, a plant will produce floral primordium through the formation of hypothetical substances. These substances are 'vernalin', if stimulated by proper temperature, and 'florigen', if stimulated by proper photoperiod. This floral primordium will ultimately developed into flower, which bloom sooner or later.

Factors Affecting Flowering

Temperature

Temperature, especially low temperature, is an important factor for flowering. Winter wheat, for example, will not flower at all if not exposed to low temperature; likewise, colored bracts underneath poinsettia flowers will not turn red or other colors if the temperature is high.

Temperature is particularly effective in flower initiation of bulbs, especially the type in which flower primordia are formed during storage after harvest in the summer but before replanting in the fall. Stored bulbs, e.g. tulips, irises, can be accelerated to flower by exposure to low temperatures (9-13°C), but a high temperature (20 to 30°C) pretreatment is essential if flower formation is to occur at all.

Photoperiod

This is a cycle of day length within a 24-hr period. Flowering of a plant in response to photoperiod is known as 'photoperiodism'. Plants can be classified as a short-day, long-day, or day-neutral plant. A short-day plant is the one in which flowering takes place when the day length is shorter than a critical value (any number of hours of day length per day); while a long-day plant is the one in which flowering takes place when the day length is longer than a critical value. A day-neutral plant is the one in which flowering is not affected by photoperiod.

Flowering Behavior

Depending on their genetic constitutions, plants can be classified into two groups, namely:

(i) those that flower all year round, and

(ii) those that flower only in-season.

All-year round: There are two types of plants in this group based on how much the season can influence, namely:

(i) *Little or no seasonal influenced*: Flowering occurs all year round with little or no influence of the season. Examples are roses, marigold, chrysanthemum, and heliconia.

(ii) *Great seasonal influenced*: The season greatly influences flowering in this type of plant. There is a time when flowering occurs profusely due to favorable climatic conditions, and at the other times, not quite profusely, simply because the weather is not optimum. Examples are jasmine, Dendrobium orchids, etc.

Seasonal: These are plants that flower during specific seasons. There are two types of plants in this category, namely:

(i) *Temperature Influenced*: These are plants whose flowering is influenced by the temperature, especially low temperature. Examples are bulbs including amaryllis, tulip, daffodil, narcissus, etc. which will flower upon exposure to low temperatures.

(ii) *Photoperiodic Influenced*: These are plants whose flowering is influenced by the photoperiod. They are divided into two kinds:

(a) *Short-day plants*: These are plants that flower when the day length is shorter than a critical value. Temperature may have influence on flowering; e.g. poinsettia, in which the bracts will not turn red if the temperature is high, although flowers are borne after exposure to short days.

(b) *Long-day plants*: These are plants that flower when the day length is longer than a critical value. Humidity may have influence on flowering; e.g. Siam tulip, in which the initiated flowers will not develop if the humidity is low.

Objectives of Flower Forcing

The objectives of forcing a plant to flower during off-season or at certain specific dates are as follows:

1. *To avoid surplus of in-season cut flowers*: Most cut flowers are produced during their respective seasons that have favorable conditions suitable for their production. Thus, large quantities are available and, as a consequence, fetch a low price or even un-saleable.

2. *To avoid wastage or spoilage of surplus cut flowers*: Unlike most other products, cut flowers are perishable with time. They are either spoiled or left as waste if not used, or sold.
3. *To avoid danger of epidemics*: Inseason production of cut flower is subject to a number of insect attacks and disease epidemics due to the presence of favorable climatic conditions for their growth.
4. *To distribute employment throughout the year*: Cut flower production requires intensive labor. During the normal peak season of production, most laborers are engaged in one job or another. Off-season cut flower production helps to distribute employment throughout the year; a benefit to the nation's economy.
5. *To increase farmers' income*: It is obvious that the farmer's income will be higher by selling products that are on high demand.
6. *To reduce imports and balance of trade*: It is the florists' custom to import cut flowers from abroad to satisfy the demands of the customers during certain times of the year when domestic production is not possible, even at a much higher price. Off-season cut flowers, or those produced at certain specific periods, help to reduce imports and balance of trade.
7. *To satisfy the customers at the time of needs*: Generally speaking, the need for cut flowers of the customers is not seasonal, but spread all-year round depending on certain specific occasions. Off-season cut flowers, or those produced at certain specific dates, are to satisfy the customers at the time of their needs.

Flower Forcing Operation

Adjusting Factors Affecting Flowering

Temperature

For low temperature-requiring plants, the temperature can be adjusted by keeping the plant parts, mostly bulbs or seeds, in the refrigerator or freezer.

Photoperiod

Photoperiod can be adjusted either by providing supplementary light from artificial sources, such as tungsten bulbs and fluorescence lamps, to prolonging the duration of day length, or keeping the plants in a dark room to shorten the duration of day length. There are two main aims of adjusting photoperiod:

— *To induce flowering by giving optimum photoperiod*: By providing supplementary light to a long-day plant to exceed the critical value, flowering will be induced; e.g. peppermint requires at least 16 hrs of day length to flower, whereas the maximum day length in Thailand is 14.30 hrs; thus a little over 1.30 hrs of supplementary light is needed to induce peppermint to flower in Thailand.

— *To keep the plant in noninductive cycle so that it will not flower:* By providing supplementary light to a short-day plant to exceed the critical value, it will not flower, but remains in the vegetative phase; e.g. chrysanthemum, a short-day plant, requires 14.30 hrs or less to flower, whereas the day length in Thailand is less than that, thus it will flower all-year round even from a small plant, making it non-productive, and only small flowers are produced. By providing supplementary light to exceed 14.30

hrs of day length, it will not flower, until such time that is optimum, light is cut off, and it will flower profusely.

Chemical Flower Forcing

Four types of chemicals affect flowering, namely fertilisers, plant hormones, ethylene, and other chemicals.

Fertilisers

Certain fertilisers affect the C/N ratio of the plant, which in turn affects flowering. The broader C/N ratio, i.e. higher C, will induce flowering, while the narrower C/N ratio, i.e. lower C, will keep the plant in vegetative phase. Adjusting fertiliser formula can be used to retard or stimulate flowering.

Retarding flowering

This can be done by providing fertilisers having high amounts of N to the plant. Watering should also be provided so that N will be readily absorbed by the plant.

Stimulating flowering

This can be done by giving fertilisers having low amounts of N, and reducing watering. Other chemicals that help to fix N to a bound form can also stimulate flowering.

Plant hormones

There are two main types of plant hormones what affect flowering, namely:

— *Gibberellins*: At least 50 gibberellins have been discovered in fungi and plants. All could properly be called gibberellic acids, or GA. Gibberellins have the unique ability among plant hormones to stimulate extensive growth of intact plants. It has been demonstrated that gibberellins can substitute for the long-day requirement in some species, and has an interaction with light. They also overcome the need some species have for inductive cold period to flower (vernalisation). It appears that the formation of flowers caused either by long days or by cold periods might normally depend upon the buildup of endogenous gibberellins during these periods, because the gibberellin content on some affected plants increase following these treatments.

— *Growth retardants:* These are a group of synthetic chemicals that inhibit stem elongation and cause overall stunting. They do so in part because they inhibit gibberellin synthesis. These include: Phosphon D, Amo-1618, CCC or Cycocel, and Ancymidol. Growth retardants promote initiation of floral primordium by reducing endogenous GA level, or antagonising its inhibitory effect on floral initiation.

— *Ethylene*: Ethylene has been popularly known to induce flowering in the pineapple. A sprinkle of acetylene, a precursor of ethylene, on top of the pineapple plant, is quite effective in inducing the pineapple plant to flower. An ethylene-releasing substance, called ethephon or Ethrel, is commercially available. The induction of flowering in the mangoes and the bromeliads by ethylene is unusual, because the gas inhibits flowering in most other species.

— *Other chemicals:* There are a number of other chemicals that are used to induce flowering, especially of fruit trees. These include the explosive potassium chlorate and its related compound -

sodium chlorate, potassium nitrate, thiourea, paclobutrazol (Cultar), etc. which, when applied as soil drench or leaf spray results in flowering of many fruit trees.

Mechanical Flower Forcing

Pruning

Pruning helps to broaden the C/N ratio, thus stimulating flowering. This can be seen in the case of bougainvilleas in which flowering takes place soon after pruning, with the application of proper fertilisers and watering. Other flowers, such as roses, also need pruning for flowering induction.

Leaf Trimming

Leaves may provide inhibitor to flowering, like in the case of jasmine. Thus, cutting off some leaves or portion of a leaf helps to induce flowering.

Ringing

Ringing helps to broaden the C/N ration, thus stimulating flowering. This is obvious in the case of fruit trees in which ringing induces flowering and fruiting.

Low-temperature storage

Many plants are promoted to flower by lowtemperature storage. This can be done by keeping the plants, usually the bulbs or corms, in the refrigerator for a period of time before they are taken out for planting.

Breaking dormancy

Seeds and buds of some plants are in dormant stage, i.e. no growth at all, for a period of time. Breaking dormancy can be done by exposure to low temperature, or treatment with chemicals and gibberellins. The latter, i.e. gibberellins, are more commonly used in association with flowering. Applied gibberellins break dormancy of many cold-requiring seeds and induce flowering of many cold-requiring plants.

Forcing Operation in Common Cut Flower Plants

Flower forcing is an operation or treatment to the plant, after it reaches the ripeness-to-flower stage, in order to stimulate it to flower at a specific date, or during off-season period. The flowering date/period may be earlier or later than the normal date/period of flowering. The goals of flower forcing are off-season production and specific-date production.

Cut flowers which are available during the normal season are plentiful, thus fetching a low price. Sometimes the farmers have to sell their produce even at a loss. In some cases, flowers which could not be sold are either left on the plants or are spoiled after being harvested. Thus, it would be beneficial for farmers to produce cut flowers during the off-season period to obtain higher price, although the inputs may be higher.

Forsythia

The first step in forcing flowers is to determine which plants are good candidates for the procedure. It is certain woody plants that are susceptible to forcing. Forsythia bushes and pussy willows are two of the best candidates — and two of the most widely available plants. The instructions below show you how to force flowers on forsythia stems.

The second step is to learn when to force flowers. February and March are the months to force flowers, since by February the plants have endured enough cold weather to satisfy their chilling requirements. After March, it would be rather pointless to force flowers from forsythia, since, by then, they're ready to burst forth outdoors, naturally.

With these preliminary considerations out of the way, now we can move on to the fun part of forcing flowers: namely, going out to cut the branches. This may become habit-forming. Once you do it, it will become an annual rite of spring for you.

The only equipment you need is a sharp pair of pruning shears. Here's how to force forsythia flowers:

— Pick a day that is above freesing.

— Cut the forsythia stems in lengths of less than 3 feet, and bring them home.

— Put the stems in a bucket of warm water.

— With your pruning shears, cut another inch off the bottoms of the submerged stems. This second cut, performed underwater where air cannot act as a drying agent, will promote water intake.

— Allow the forsythia stems to soak up the warm water for several hours.

— Now it's time to change the water. When you refill, put floral preservative in the warm water this time. Once again, re-cut the stems, underwater. You'll speed up the forsythia flower forcing if you keep the stems in a high-humidity environment and give them some sun.

Bulbs

Making a plant flower at a predetermined time or under artificially imposed conditions is called forcing. Hardy bulbs are particularly suited for forcing indoors and offer a succession of color throughout the winter and spring months.

The most common hardy bulbs for forcing are crocuses (Crocus species), daffodils (Narcissus species), hyacinths (Hyacinthus species) and tulips (Tulipa species). Others that can easily be forced include Dutch iris (I. x hollandica) and netted iris (Iris reticulata), snowdrop (Galanthus species), grape hyacinth (Muscari species), winter aconite (Eranthis species), star-of-Bethlehem (Ornithogalum species), and Brodiaea species.

Allium, Camassia, Lilium and Scilla species are considered difficult-to-force bulbs; they require special techniques and usually greenhouse conditions (e.g., high light intensities).

Forcing hardy bulbs involves four stages:

(1) selecting appropriate bulbs;

(2) planting;

(3) cooling; and

(4) forcing into flower.

Selecting appropriate bulbs

For the best results, purchase only those cultivars that are recommended for forcing, especially if a wide number of cultivars are available, as is the case with tulips, hyacinths and daffodils. Handle the bulbs with care at all times. They are living plants and should not be dropped or subjected to extremely low or high temperatures. If you cannot plant your bulbs immediately, store them in a cool place (35 to 55°F). Bare bulbs can be stored for several weeks in the refrigerator prior to potting. Store them in a mesh bag or a paper bag with holes to permit ventilation. Vegetable or crisper drawers can be used, but avoid storing bulbs in the same drawer as ripening fruit or vegetables which give off ethylene gas which may harm the bulbs. Also, some bulbs are poisonous, so this storage method is not recommended for households with young children.

Planting

Bulbs should be potted up anytime from mid-September to December, depending on the desired date of flowering and the length of storage. In general, plant in mid-September for flowering in late December, around mid-October for flowers in February and in mid-November for March and April flowers.

The potting medium must be well-drained, retain adequate moisture and be able to anchor the bulbs. Commercial potting soil is adequate, but a better choice is a mix composed of equal volumes of potting soil, sphagnum peat moss and perlite. Since the bulbs already contain enough food for the developing flowers and roots, they do not have to be fertilised. Use clean pots that have drainage holes in the bottom. If you use clay pots, soak them overnight so they won't draw moisture from the planting medium.

Fill each pot loosely with soil. Ideally, bulbs should be planted at the same depth as bulbs grown outdoors; however, this is often not possible with larger bulbs which need a pot that is deep enough to allow at least 1 to 2 inches of soil beneath the bulb. Tulips and daffodils may be left with the tips of the bulbs showing; smaller bulbs such as crocus, snowdrop and grape hyacinth should be covered completely. Do not press the bulbs into the soil; the soil should be loose so roots can grow through it easily. A 6-inch pot will accommodate six tulips, three hyacinths, six daffodils, or 15 crocuses. When planting tulips, face the flat side of the bulb toward the outside of the pot. Since the first (lowest) leaf produced by the flowering shoot is always produced towards the flat side of the bulb, the lowest leaves will arch over the rim and create an attractive, uniform appearance. After setting the bulbs, fill the pot with soil to within ¼- to ½-inch of the rim. Add water until it drips through the drainage hole in the bottom of the pot. Label each pot, noting the name of the cultivar, the planting date and the date to be brought indoors for forcing.

Cooling the bulbs

All hardy bulbs must be exposed to cool temperatures between 35 and 50 °F to prepare them for subsequent leaf and flower growth. During this cool period, the plant's roots are forming and the stem is elongating. The optimum cooling period is from 12 to 16 weeks. Although they can be cooled anywhere from 12 to 18 weeks, the stem length will suffer, becoming shorter (flowers may also abort) if the bulbs have been cooled for fewer than 14 weeks and longer if the bulbs have been cooled for more than 15 weeks. If the bulbs were held in the refrigerator for more than three weeks, subtract three weeks from the required cooling time.

Any structure that maintains temperatures between 35 and 50 °F is fine (e.g., an unheated basement, crawl space, or an old refrigerator). For areas that experience sufficiently cold temperatures during the winter months, pots can be stored outdoors in a cold frame. Locate the cold frame in a well-drained, shady location or on the north side of a building where the soil is as uniformly cool as possible. Place the pots in the coldframe and cover them with loose insulating material such as sawdust, straw, leaves, peat moss or shredded styrofoam. Close the frame early in December and open it on mild days to prevent too much heat from building up inside the frame, stimulating top growth before the roots have formed. Check the pots periodically to see that the medium is moist.

Forcing blooms

At the forcing stage, the pots are taken out of storage at the completion of their cool period and into warmth and light, which triggers the formation of leaves and flowers. Place the pots in a cool location (60 to 65 °F) receiving indirect sunlight. When the shoots turn green, expose the pots to warmer temperatures and more light to stimulate growth and flowering. Rotate the pots regularly so that all the leaves receive an equal amount of light. Flower buds can be expected within three to four weeks. When the buds begin to show their color, move the pots into indirect sunlight to prolong the flowers. Be sure to keep the soil evenly moist throughout the forcing period. Flowers will last longer if the pots are moved into a cool room at night.

Hardy bulbs that have been forced into flower should be planted in the garden once spring arrives, or allowed to mature and go dormant in their pots and then planted in the fall. Daffodils can be transplanted into the garden in the spring. They will not flower the following year, but may the year after. Other bulbs such as tulips and hyacinths are best discarded after forcing.

Forcing without cooling

The paper white narcissus (*N. tazetta* 'Paper White'), its yellow cultivar Soleil d'Or, and the Chinese sacred lily

(*N. tazetta* var. orientalis) can all be forced without cooling. Successive plantings made about two weeks apart after mid-October can produce indoor flowers from Thanksgiving until late March.

Start by filling an undrained decorative bowl or dish that is at least 2 to 3 inches deep with enough pebbles, pea gravel, coarse sand or pearl chips to reach about 1 inch below the top. Add water until it is barely below the surface of the gravel. Set the bulbs on top and hold in place with enough gravel to cover the bottom quarter of each bulb. Carefully maintain that water level.

Tender Narcissus are best kept in a cool 50 to 60 °F location in low light until they are well-rooted and the shoots appear, usually in about two to three weeks. Then bring them gradually into direct sunlight and warmer temperatures. These bulbs cannot be forced again after blooming; discard them.

Amaryllis (Hippeastrum cultivars) can be forced inside in four to six weeks without cooling, and then planted outside in the spring or allowed to remain in pot culture. To program potted amaryllis for forcing, withhold water in mid-August, then start watering again about six weeks before you want them to flower.

Marigold

Flowers are available round the year. It is a day neutral plant which takes 60 - 70 days from seeding to harvest. Timing of flowering can be maintained by fixing the date of seeding 60 to 70 days prior of harvest date. The recommended period is 65 days. For example, if the flowers are to be harvested for New Year, seeding should be done on 27 October, transplant on 6 November and pinching on 22 - 24 November. Flower buds emerge by 5 December, and blooming starts from 25 December to 5 January.

Lotus

Flowers are available all the year round, but the plant needs standing water. Low temperature during winter reduces the amount of blooming. Thus, it is recommended that the level of water be reduced to 50 cm in order to raise water temperature. In this way, blooming will be the same as in summer, both in terms of amount and size of flower.

High temperature during the summer speeds up the growth of lotus plant and the blooming time, but flower size remains small. Thus, water level should be raised up to 75 cm. This way, water temperature will be reduced and the amount of bloom and size of flower will be maintained. During the rainy season, due to the addition of rain water the level of water may increase, thus it is recommended that the level of water be maintained at 50 cm in order to accommodate additional rain water. The blooming will be maintained as normal.

Chrysanthemum

It is a short day plant, with critical value of 14.4 hours. Thus, it will bloom all the year round under Thai conditions, having maximum daylength of 13.3 hours in June. Daylength can be extended by giving artifical light after sunset for about 3 hours during the early stage of growth to keep seedlings in vegetative stage until one month prior of the planned harvest date. For example, if the planned harvest date is New Year, cutting should be made in September and transplanted to the growing plot for rooting to occur. Seedling is kept under light regime of more than 14.5 hrs by providing artifical light (100 w incandescent bulb) until 1 December (seedlings should be at least 30 cm high). These will bloom on 1 January.

However, as chrysanthemum blooms profusely during the period of low temperature which commences in December, it fetches a low price in the market even during the time of Christmas or New Year. Thus, some farmers avoid producing flowers during such period but shift it to the summer. The problem is that the temperature during summer is quite high and not optimum for chrysanthemum growth.

The most suitable place of production of chrysanthemum flowers is on the highlands of northern Thailand where the temperature during the summer is around 16-20°C. The same principle of flower forcing is applied, but in this case the daylength in northern Thailand may be lower than the critical value for certain varieties. As practiced in Doi Phu Kha of Nan Province, black cloth is used to completely cover the plant house from 16.00 to 08.00 hrs. for 30 days after the cuttings have been exposed to long day conditions.

Jasmine

Flowers are available thoughout the year, but very profuse during the rainy season and scarce during the winter. To produce jasmine for winter-season harvest, the following operations are recommended:

i) One month before the planned date of harvest, stop watering for 2 - 3 days until the plants show sign of wilting.

ii) Prune the plant to a round shape so that blooming will emerge from mature branches when induced in stage

iii) In this way the blooms will be of large size and healthy as they receive full sunlight.

iii) Apply balanced fertiliser at the rate of 30 g/plant and water heavily. Keep watering normally everyday. Flower buds will emerge within 10 days and blooming will occur within 25-30 days after pruning.

Gladiolus

Flowers can be available all the year round, provided the weather is optimum (requires cool climate). Blooms 90 - 100 days after planting. In cool climate preheat corms before planting for 2 weeks at 27-32°C. This will force such corms to flower early. In warm climate soak in GA3 solution (10 - 25 ppm) before planting. This will accelerate flower by hastening differentiation of floral primordia.

Growth retardants (e.g. CCC) promote initiation of floral primordia (by reducing endogenous GA3 level, or counterbalancing its inhibitory effect on floral initiation.

Roses

Flowers are available round the year but more profuse in the cool season. For blooming during Christmas - New Year, and Valentine Day, the following procedure is suggested:

i) Cut the branches in November. This will stimulate flowering during Christmas - New Year. Flowering will take place 43 days after this pruning. Cutting the flowers on 23 December will further stimulate flower bud initiation to bloom on 10 February. Flowering will take place 49 days after this cutting.

ii) Pinch (i.e. cut only the tip off) instead of pruning as in (i) above, on 10 November for 23 December harvest. This operation automatically stimulates flower initiation for 10 February harvest.

The following operations should be followed:

- Do not pinch healthy flower buds after 10 November as blooming will be due in 30 days. Allow them to bloom for 10 December harvest.
- Do not pinch flower buds which are still very small.
- Pinch all other branches, including those flower buds which are not healthy, deformed, etc. Do not pinch too far down to healthy leaves, unless they are water sprouts which should be cut off from the base.

The technology of flower forcing has been accumu-lated and placed on a trial-and-error basis for a few cut flowers. However, due to erratic climatic conditions, such practices will have to be modified to avoid the risk of obtaining flowers prior to, or after the required period. Another important factor influencing the production of cut flower availability at certain specific times or off-season is the use of chemicals. As happens in forcing off-season fruits like longan, mango, durian, etc., the use of chemicals has been blamed by the environmentalists. The reason given is that such a treatment creates pollution in the soil and water, as well as deteriorating the physical conditions of the soil, if such chemicals are used in high concentrations and/or for prolonged periods. Thus, the use of chemicals should be minimised as much as possible, particularly those that are used on the soil.

6

Propagation of Flowering Plants

Plant propagation is the process of artificially or naturally propagating (distributing or spreading) plants. Most plants reproduce more of their kind through production of seeds. This is sexual propagation and it involves the exchange of genetic material between two parent plants. Many ornamental plants do not come true from seed. To increase the numbers of these plants, gardeners and horticulturists use asexual propagation. In asexual propagation, the new plants are genetically exact copies or clones of a single parent plant. The methods used in asexual propagation range from taking leaf cuttings of African violets to grafting apple cuttings onto root stocks.

Sexual Propagation

Sexual propagation of plants involves the exchange of genetic material between parents to produce a new generation. Seeds and spores can be used for reproduction. Seeds are typically produced from sexual reproduction within a species, since genetic recombination has occurred plants grown from seed may have different characteristics to its parents. Some species produce seed that requires special conditions to germinate like cold treatment. The seed of many Australian plants and plants from southern Africa and the American west require smoke or fire to germinate. Some plant species, including many trees do not produce seed until they reach maturity, which may take many years. Seed can be difficult to acquire and some plants do not produce seed at all.

Sexual propagation offers the following advantages:

— It is usually the only method of producing new varieties or cultivars.

— It is often the cheapest and easiest method of producing large numbers of plants.

— It can be a way to avoid certain diseases.

— It may be the only way to propagate some species.

Propagation by Seed

Propagation by seed is the most commonly known method of producing new plants. Common annual and biennial vegetables and flowers are grown easily from seed. Perennials grown from seed may take more than one season to flower. Lawn grasses are commonly grown from seed by the home gardener.

Seed purchasing

Use seed that is guaranteed reliable for freshness and purity. Purchase seed that is packaged for the current year. Most seed companies provide information on expected percent of germination. This figure tells what percent of seed is expected to sprout or germinate. Generally, 65-80% of most fresh seed sown will germinate. Of those seeds that germinate, 60-75% will grow into satisfactory seedlings.

When choosing seed, be sure to select varieties or cultivars that meet your requirements for size, colour and growth habit. Choose varieties that will mature before frost. This can be estimated by using the number of days to maturity (found on the seed packet or in the catalog) and comparing it to the days between probable planting date and the average frost date of the area. When considering seed not packaged for the current year, remember that the germination rate of most seeds drops dramatically after the first year.

Seed catalogs are not only helpful in purchasing seed, but they are also excellent references for cultural information. Catalogs and seed packets may provide information on country of origin, bloom time, specific germination requirements, cultural requirements and disease resistance. Some seed packets indicate if seeds have been chemically treated to prevent disease. Read packets and catalogs thoroughly and follow their recommendations. The more information the gardener possesses, the greater the chance of growing plants successfully.

Seed collection

Seed saved by the home gardener will probably be the result of random pollination by insects or other natural mechanisms. Random pollination results in seeds that produce plants that may not be identical to the parent plant. The seeds of hybrid cultivars should not be saved.

Some plants make excellent candidates for seed saving. Common self-pollinated, non-hybrid and purebred annual vegetable seeds that can be saved include lettuce, beans, peas, herbs and heirloom tomatoes.

Saving seed saves money. It allows the gardener to maintain varieties that are not sold commercially. Many avid seed savers belong to groups that exchange seed through networks. Some seed saver groups specialise in keeping heirloom varieties. Many heirloom varieties are the great-grandparent plants of modern cultivars.

It may be tempting to bring home seeds or plants seen on vacation in foreign countries. However, this is how many serious insect and disease pests are introduced. A nonnative plant may become a noxious weed. Follow all import regulations for horticultural materials.

Seed harvesting

It is important to save seed from healthy plants because some diseases can be carried in seeds. Commercially grown seed is protected from disease problems because it is produced under very strict conditions with frequent inspection.

Harvest seed just before fruit is fully ripe. For flowers with exposed seeds, place the seed stalk or flower head in a bag and store in a warm, dry location. Seed will fall into the bag when it is completely dry. The seed of pulpy fruits should be separated from the pulp, washed and thoroughly dried.

Seed storing

Once seeds are completely dry, place them in airtight storage containers marked with name and date saved. Store seeds at 40 degrees F with low humidity. The refrigerator provides these conditions.

Seed of many plants can remain viable for up to 5 years if properly stored. However, it is best to use home-harvested seed during the following growing season. Some species of plants produce seeds that are short-lived. These seeds must germinate immediately after they ripen or they lose their viability. Delphinium, onion and parsley are examples.

Before planting, it is a good idea to check stored seed for its germination rate. Planting these seeds directly in the garden may be a waste of time and effort if germination rate is very low. To check germination rate, place some of the seeds between paper towels that are kept constantly moist and between 65 and 70 degrees F. Check the seeds daily for germination. If the germination rate is 70% or less, consider buying new seed.

Seed germination

When germinating seed it is helpful to remember that a seed is made up of three parts:

—an outer protective coat

—a food supply under the seed coat (the endosperm)

—an embryo of a young plant

The protective coat prevents sprouting until ideal growing conditions exist. Bringing seeds out of dormancy involves manipulating conditions to hasten germination. Even with ideal conditions, some seeds are still very difficult to germinate.

There are several factors that affect germination. Water (moisture), light (or dark), oxygen and heat play a part in triggering germination. In addition to environmental factors, seed must be viable.

Water is essential in the first phase of germination. Water penetrates the seed coat and causes the endosperm to swell. The seed coat, softened by water, splits open as the endosperm swells. The water dissolves nutrients in the endosperm making them available to the embryo and growth begins.

The growing medium must be constantly moist, but not wet. Any dry period may cause death of the sprouting embryo.

Light can stimulate or inhibit a seed's germination. This determines whether the seed should be sown on the surface of the growing medium or below the surface. Check the seed packet or catalog for light requirements.

Oxygen is required by the embryo to begin growing. The seed must respire to break down the food stored in the seed. This is one reason for using a light, well-aerated growing medium for starting seeds.

Every seed has an optimum temperature range for germination. Many seeds have a fairly wide temperature range for germination, but some are limited to a narrow range. The temperature range is usually given on the seed packet or in the catalog.

The temperatures required by many seeds are higher than those in most homes. The desired constant temperature can be achieved through heating cables placed under germination containers. Setting flats or pots on radiators, the furnace or on the refrigerator will provide bottom heat. However, these locations may be too hot and cause the soil to dry too quickly.

Once germination occurs, a different, usually lower, temperature may be required for optimal growth of the seedlings.

Seed scarification

The coat of certain seed is extremely tough and must be penetrated by special means. Particularly hard seed may be scarified. Scarification involves breaking, scratching or softening the seed coat to allow moisture penetration.

Two methods of scarification commonly used by the home gardener are mechanical and hot water. Mechanical scarification involves breaking or weakening the seed coat with a file, sandpaper or hammer. Hot water scarification involves placing seeds in water that is 170 to 210 degrees F. After the water cools, seeds should continue to soak for 12 to 24 hours. Then they are planted. Specific instructions for scarification are usually mentioned on the seed packet or in the seed catalog.

Seed stratification

Some seeds will not break their dormancy unless exposed to a period of low temperature and moist conditions. Stratification requirements are usually indicated by the seed supplier.

Plants that typically require seed stratification include many trees and shrubs, and certain perennials. This period of "chilling" or stratification can be accomplished by placing seed in a moist, sterile (pasteurised) growing medium, such as a mix of equal parts clean sand and peat or sphagnum peat moss, in a disinfested container. Enclose the container in a tightly sealed plastic bag and place it in the non-freezer section of the refrigerator. Towards the end of the stratification period seeds may start to sprout. Handle sprouted seeds carefully when transplanting into pots. Some seeds may require both scarification and stratification to germinate reliably.

Sowing seed Indoors

Sowing seed indoors is the easiest and cheapest method of producing vegetables, annual flowers and some perennial plants. Plants with extremely small seeds or those that need a long growing season make excellent candidates for starting indoors.

Very small seed is difficult to plant outdoors because of size. Plants that require a long growing season may not have enough time to reach maturity unless started indoors in winter or early spring.

Supplies needed for indoor seed sowing include the following: fluorescent or grow lights, disinfested containers with excellent drainage, pasteurised (sterile) seed-starting medium and a location with proper temperature and ventilation.

Growing media. Choose a medium with a loose, uniform, fine texture. A pasteurised mixture that is 1/3 soil, 1/3 sand, vermiculite or perlite, and 1/3 peat moss has the qualities of a good seed-starting medium. Retail garden centers carry mixes labeled for seed starting. Whatever is selected, be sure it is pasteurised (sterile). Using pasteurised soil prevents damping-off, a fungal disease that kills young seedlings. Pasteurised soil also helps to avoid weeds, diseases and pests. Seed-starting media are usually low in fertility. This means that a regular fertilisation programme is very important once seedlings emerge.

Containers. Any recycled containers are adequate for seed starting provided they are disinfested, have good drainage and are at least 2 inches deep. Other container options include compressed peat pellets, peat pots, paper pots, plastic cell packs and flats. Peat and clay containers tend to dry more quickly than plastic containers because they are very porous.

Seed sowing. The correct timing of seed sowing is an important factor in successful indoor seed starting. In winter months, overanxious gardeners may sow seeds too soon. Seedlings that are held indoors too long perform poorly once transplanted into the garden. Most seeds should be sown 4 to 12 weeks prior to transplanting into the garden. The time it takes for seedlings to be ready for transplanting outdoors will vary.

An acclaimation period before placing seedlings directly into the permanent growing site must be included. However, readiness for outdoor planting will vary with how quickly germination occurs, the growth rate and weather conditions. Quality and quantity of light, temperature and nutrients affect the growth rate. Seed catalogs and packets provide information on days to germination and weeks needed to reach transplant size.

Fill the container to within 1/4 inch of the top of the container with moistened seed-starting medium. After sowing seeds, keep the medium moist, not wet. To keep the medium moist, you may place the container in a plastic bag just large enough for the container. Seal the bag. The plastic bag keeps moisture in, but allows air exchange. The plastic bag method should not require any further watering until germination. Provide proper light and temperature conditions.

Once seedlings germinate, remove the container from the plastic bag. Place the container in a location that has high light intensity and cooler temperatures.

Use the following rules to sow seed, depending upon seed size. Sow very small seeds by sprinkling on top of the medium and pressing in. Use a fine mist of water to gently wash seed into the growing medium. Sow medium-size and larger seed in rows 1 to 2 inches apart, and 1/8 to 1/4 inches deep. If no depth is specified on the seed packet, use the general rule of planting the seed at a depth twice the diameter of the seed.

When sowing in a tray or flat, sowing in rows is preferred over scattering seed. This method provides better air circulation than scattering of seed. When seedlings are crowded, they may become tall and spindly. To avoid the need to transplant seedlings from a seed flat to pots, you may sow seeds directly into cell packs or peat pots. Plant two or three seeds per cell or pot. When they germinate, remove the two less vigorous seedlings.

Transplanting. Once the seeds have germinated, you will see two seed leaves or cotyledons. Eventually, these seed leaves will wither. Next, true leaves will form. These true leave look like the plant's typical leaves. Transplant the seedlings to individual containers once the first set of true leaves appear.

Use a pencil, small stick or other narrow tool to lift seedlings from the seed flat. Plant each seedling in its own small pot filled with pasteurised growing medium. The new container should be no larger than a 2-1/2- to 3-inch pot.

Hold the seedlings by a leaf between the thumb and forefinger. Handling by the leaf avoids damaging the fragile stem which would kill the seedling. If a small leaf is lost in transplanting, the plant can recover and grow normally. Expose seedlings' roots to air as briefly as possible. Lightly firm the soil around the seedling.

Transplanting temporarily slows or stops the growth of seedlings. Sowing into cell packs bypasses the need to transplant the seedlings.

Watering. Watering can be a cause of seedling failure. Keep soil moist but not wet. Small, tender seedlings dry out rapidly and can die. Remember that roots always must have oxygen, as well as water; therefore, do not keep the soil soaked. Water when the surface of the soil begins to dry out.

Bottom watering helps prevent damage to the seedlings caused by a hard stream of water. Bottom watering also encourages deep root development and ensures that the entire depth of soil receives moisture. Do not let the pot or flat sit in water longer than it takes for all of the soil to become moist.

Temperature for seedling indoors. Keep seedlings in a well-ventilated, cool location. The temperatures should be 55 to 60 degrees F at night and 65 to 70 degrees F during the day. These temperatures encourage compact, bushy, vigorous growth while minimising disease.

Light requirements. Seedlings require bright light immediately after germination. One warm-white, 40-watt bulb and one cool-white, 40-watt bulb used together are adequate for seed starting and seedling growth. Fluorescent lights can be used for one year before replacement is recommended.

Special grow lights are also suitable, but more expensive. The lights should be no more than 6 inches above the top of the seedlings. Mount the light fixture so it can be raised as the plants grow in height.

Day-length requirements vary with different plants. Petunia, snapdragon, phlox and China aster require short daylight periods of 10 to 12 hours. However, most plants that are started from seed benefit from 16 to 18 hours of light.

Ferlilisation of Indoor Seedlings. Because the growing medium used to start seedlings is usually low in nutrients, a regular fertilisation programme is important for proper plant growth. Apply a liquid fertiliser high in phosphorous weekly. Fertiliser with a 1-2-1, N-P-K ratio is recommended. Dilute fertiliser 1/4 to 1/2 the label's recommended strength and apply sparingly. Always use a liquid form of fertiliser.

Pinching of seedlings. Pinching the growing tips of seedlings will result in more branching. This produces a fuller, stockier plant.

Hardening off Seedlings before Transplaning. Plants grown indoors must be gradually introduced to outdoor conditions. If seedlings are not hardened off, leaves may be burned by the intensity of the sun or damaged by wind. Acclimate plants by first placing them in a cool, protected location, such as a porch or shaded COLD FRAME.

This first step in hardening off allows plants to adjust to outdoor temperatures. After 7 to 10 days, move seedlings into a shaded area of the garden for 2 to 3 days. This will prevent sunscald. Finally, hardened seedlings can be planted directly into the garden as weather permits. Planting on a cloudy day or late in the evening is a sensible precaution.

Transplaning Seelding into the Garden. The garden soil should be adequately dry to prevent compaction. At this point, plants will again experience transplant shock and a setback in growth. Plants must adjust to dramatically different nutrient levels, soil temperatures, moisture levels and soil tilth in the garden.

Pull apart the lower portion of the root mass to get the roots growing outward. If seedlings have been grown in peat pots, pull apart the bottom of the pot and roots. Although seedlings may be planted without removing the pot, be sure to maintain the same soil level. Trim away any of the pot that is above the soil line. The exposed portion of the peat pot acts as a wick and dries out the entire pot and roots.

Water seedlings into the soil. A cup of transplanting solution will help plants get off to a good start. Make your own transplanting solution by mixing 1 tablespoons of a water-soluble 20-20-20 fertiliser in a gallon of water.

Sowing Seed Directly into the garden

Many flowers and vegetables may be sown directly into the garden. Direct sowing avoids transplant shock. It takes less work but involves more risk from weather, pests, diseases and erosion.

Before sowing seeds directly into the garden, know what conditions are required for germination and growth. A warm-season crop, such as beans, may rot before germinating if planted in cold, damp soil. Knowing the average frost date for your area helps to avoid losing frost-sensitive plants. Some cool- season crops, such as peas and lettuce, should be planted early in the season while temperatures are low.

Sow seeds in a row or broadcast them into a well-raked seedbed. The seedbed should be free of stones or other large debris. Choose a calm day. To broadcast seeds, merely scatter them over a large area in the seedbed. Cover the seeds with a fine layer of soil. To sow very small seeds, mix them with sand before scattering. Then water with a gentle spray. Avoid washing seed away when watering.

Propagation by Spores

Ferns can be propagated from spores which develop in clusters on the underside of fronds. Collect fronds that have produced spores and store them in an envelope until dry. After drying, separate the dust-like spores from the cases by screening. Store spores in an airtight container in a cool, dry place until ready to plant.

Germinating spores requires more time and care than germinating seeds. Growing ferns from spores involves the two different generations of ferns. Spores first produce an asexual plant called a gametophyte. This plant is very small and has none of the usual plant parts. It resembles a moss-like growth and is about 1/8 inch thick.

The gametophyte reproduces sexually and forms sprophytes which have visible roots, stems and leaves. During the first phase of growing ferns, sterile conditions are critical. Moss, fungi and algae compete aggressively with young fern gametophytes. Sow spores on top of a pasteurised (sterile), moist, soilless mix or sphagnum peat in a disinfested container. Water must also be sterile. It takes from 3 to 6 months to grow ferns from spores.

Asexual Propagation

Asexual propagation does not involve exchange of genetic material, so it almost always produces plants that are identical to a single parent. Delicious apple and Bartlett pear are two examples of species that have been asexually propagated for decades. Asexual propagation methods include cuttings, layering, division, grafting, budding and tissue culture.

Cuttings

Cuttings involve removing a piece from the parent plant and that piece then regrows the lost parts or tissues. Both woody and herbaceous plants are asexually propagated by cuttings of stems, leaves and roots. New plants can be grown from parts of plants because each living plant cell contains the ability to duplicate all plant parts and functions. Mature cells can change into Meristematic cells that are found at rapid growth sites like buds. There are many types of cuttings. Often, a plant can be propagated by more than one method of cutting. Some plants will reproduce readily from cuttings and others take a considerable amount of time and care.

Stock Plants are the parent plants used in asexual propagation. Stock plants must be in excellent health and should possess characteristics desirable for production of new plants. Herbaceous cuttings are those taken from nonwoody plants, such as perennials and houseplants.

Softwood cuttings are pieces of new growth taken from woody stock plants. These cuttings must be taken before the new growth starts to harden. Hardwood cuttings are taken from tissue which has become woody. Other forms of cuttings are leaf cuttings and root cuttings.

The gardener must try to duplicate the conditions needed for a plant to root from a cutting. High humidity, indirect light and soil temperatures of 70 to 80 degrees F are best for most cuttings. These conditions may be created by keeping cuttings enclosed under glass or in plastic bags in dappled shade. Cuttings must be shielded from direct sunlight, especially if they are under glass or plastic.

Stem Cuttings

When a cutting is made, injured xylem and phloem cells plug the tubes so that precious fluids are not lost. Usually a Callus forms at the cut. Cells near the callus area reorganise to form adventitious roots.

Stem cuttings are the most commonly used method to produce houseplants. Select vigorous, new growth with no flower buds. Stem sections should be free of diseases and insects. Each cutting should be 2 to 4 inches long and have 2 or 3 leaves attached.

Make a cut 1/4 inch below a leaf node and pull off the leaves that are at the nodes that will be below the surface of the rooting medium. Rooting Hormone helps to stimulate rooting, but is optional. Pour a small amount of the rooting hormone into a clean container to prevent contamination of all of your rooting hormone. Dip the base of the stem, including the node area, into the rooting powder. The stem should be dry when dipped.

Commercial rooting products often include a fungicide. This is a good idea given the damp conditions required for rooting success. Tap off excess powder, since too much hormone can inhibit rooting.

Poke a hole in the medium before inserting the cutting to avoid loss of the rooting hormone. Insert treated cutting in a moist rooting medium. A suitable rooting medium is half perlite and half sphagnum peat moss. Any disinfested container with drainage is acceptable for use.

Cover container and cutting with a plastic bag tent to maintain high humidity. Place unit in a warm area with indirect light. Check the rooting medium every few days to make sure it remains moist. Rooting can take from a few days up to several months.

After a few weeks, test for rooting by gently tugging at the cutting. If there is resistance, rooting has started and the plastic cover may be removed.

Leaf Cuttings

In this method, a leaf blade or leaf with petiole is used to propagate new plants. The same steps are followed as for stem cuttings. Choose a healthy leaf from a vigorously growing plant. Cut it close to the stem with a sharp, disinfested razor or knife. Trim off 1/4 of the leaf and dip into rooting hormone, if desired. Insert the leaf into rooting medium so that 1/3 of the leaf is below the surface.

One or many new small plants form at the base of the leaf. African violet leaves will produce many new plants. Begonia leaves can be divided into segments for propagation; however each leaf piece must contain a major vein. With leaf cuttings, the original leaf is not a part of the new plant and is usually discarded.

Many succulent plants, such as sedum, jade, and peperomia, can be propagated by leaf cutting. Some plants such as Kalanchoe pinnata or K. bryophyllum, piggy-back or air plants, form their own new plantlets in this fashion.

Root Cuttings

Cultivation of root cuttings probably started after gardeners observed new plants growing from pieces of root accidentally left behind in the soil. Take cuttings from newer root growth. Make cuttings 1 to 4 inches long from roots that are 1/4 to 1/2 inch in diameter. Be sure that the roots collected are from the chosen plant and not neighboring plants.

Cuttings should be taken during the dormant season when roots have large carbohydrate supplies. However, they also may be taken throughout the growing season. Cut straight through the end of the root closest to the stem. Cut the other end on a slant. This allows you to remember which end is the top (the straight cut) and which is the bottom (the diagonal cut).

Store cuttings from dormant roots for 3 weeks in moist rooting medium at 40 degrees F. Remove from storage and plant upright in the growing medium. Keep moist and warm, in a bright location until growth and weather permit acclimatising to the outdoors.

If root cuttings are taken during active growth, skip the storage period and place cuttings directly in the rooting medium. For smaller plants, take 1- to 2-inch sections. Place cuttings horizontally a half inch below the surface of the rooting medium. These cuttings should be handled indoors or in a HOTBED. The fine roots of many perennials are used for propagation. Root cuttings of some variegated plants will lose their variegation.

Softwood And Hardwood Cuttings

Softwood cuttings are taken from first-year branches that have not yet become woody. Flowering shrubs are often propagated by softwood cuttings. Late spring and early summer are the best times for success with this method. Take cuttings after it rains or water is applied in the cool morning. Make a diagonal cut. The larger diagonal cut gives more area to develop roots. Keep cuttings in water before moving them into rooting medium.

Make cuttings 2 to 10 inches long. Larger cuttings produce larger plants sooner. Make cuts slightly below a leaf node. Follow the same instructions for disinfested containers, rooting medium and light and moisture conditions used by other cutting methods. Rooting hormones may be used with this method. The base of the cutting should be dry before dipping it into rooting hormone powder.

Hardwood cuttings are taken once the tissue becomes woody and the plant is dormant. Cuttings can be taken anytime from late fall after a killing frost until late winter. Select healthy wood that was produced the previous summer. Several cuttings can be made from the same branch of some shrubs.

Make cuts at a slant, 5 to 12 inches long. Basal cuts should be just below a node, while the upper cut should be slightly above a bud. Dip the basal end in rooting hormone. Mark the most terminal end of each cutting with a tag.

Bury cuttings vertically in moist vermiculite or sand. Cuttings should not freese, but must remain cool. A callus will form on the lower cut end during storage. Callus formation indicates that cuttings are ready to root.

In spring, remove the cuttings from storage. Plant in a hotbed or other protected site with morning sun exposure or filtered light. Leave 1 to 2 inches of cutting above ground. Keep cuttings moist until a root system forms. Transplant the cuttings the following spring while they are still dormant.

Layering

Layering causes roots to develop on shoots that are still attached to the parent plant. The stem is not cut from the main plant until it has rooted. Use this method when other propagation methods are unsuccessful. Layering is a good propagation choice when only a few plants are needed.

Simple layering is done by bending a branch to the ground and burying a portion of it. The tip remains uncovered. A light soil increases rooting success, as will wounding or girdling the buried section. Treatment with rooting hormone is helpful.

Hold the branch in place with a rock or peg. Layering is done in early spring while plants are still dormant or in late summer on wood that has not become woody. Plants with flexible branches are particularly suited to this method.

Other types of layering include serpentine or compound layering, continuous or trench layering, and mound or stool layering.

Divisions

Division is the cutting or breaking up of a crown or clump of suckers into segments. Each segment must have a bud and some roots. These segments are replanted and grow into new plants identical to the parent. Most perennials should be lifted and divided when they become overgrown and begin to lose vigor. Vigorous growth in most perennials occurs on the outer segments of the clump. Old growth in the center of the clump is discarded.

Carefully dig the plant, loosening the roots and lifting the plant from the soil. Split apart the main clump with two spades or forks or chop with a shovel or hatchet if the clump is firmly massed. In some cases outside segments of the plant can be removed and replanted without disturbing the rest of the plant.

A good rule of thumb is to divide fall-flowering perennials in spring and spring- and summer-flowering perennials in fall.

Some shrubs that form clumps or crowns are suitable for division. Divide in early spring while dormant. Shrubs used for division should have several clumps that are two years old or older.

Bulbs and Corms

Bulbs can be propagated by removing small bulblets or offsets that form at the base of the parent bulb. These small bulbs take 2 or 3 years to mature into plants that flower. Place offsets in rich, light soil for their development. This same procedure should be followed for plants which form from corms, such as gladiolus.

Many lilies can be multiplied by removing scales from the mature bulb. Dust the scale with a fungicide and place, base end down, in a moist growing medium in a warm, protected area. Bulblets will form at the base of the scale. In 1 to 4 years these bulblets will grow and be ready to flower.

Tubers and Rhizomes

Tuberous plants can be dug up and the tubers separated. In separating the tubers, each must have a segment of the crown that contains at least one eye or bud.

Rhizomes grow and develop buds along their length. The rhizomes can be dug and cut into sections that each contain at least one eye or bud.

Grafting

Grafting involves the joining of different segments of two different plants of the same species. A branch or bud is joined to an understock. In grafting, the cambium layers of the two different segments are aligned and grow together. Some plants graft naturally. Ivy commonly grafts itself. Natural grafting occurs where two branches are in close contact over several years.

Grafting allows gardeners to produce plants identical to a parent plant. It also allows growers to control size and shape of a tree or shrub. Apples are grafted onto root stock that regulates the ultimate size of the tree.

Grafting can produce more vigorous, earlier fruiting, or two varieties can be grown on the same tree to provide pollination as with apples. On the negative side, some grafting attempts will be rejected. This is called graft incompatibility. This might not occur in the first year, but may in successive years. Wind or winter cold can damage a graft union. Some grafted trees or plants produce large numbers of suckers. These can crowd out the desired plant or tree and are unsightly. If the graft union is planted below the soil surface, the scion will root and the rootstock will die.

Grafting is usually done in the spring. It involves collecting small branches called scion wood. These branches are 1/4 to 1/2 inch in diameter. Gather them several weeks before grafting is to occur. Collect wood that grew the preceding year. Select only wood with leaf buds, not flower buds.

Scion wood should be gathered in winter when wood is dormant, but not frozen. Use sharp clippers so that wood is not crushed. New growth over 1 foot in length is usually best.

Discard the wood at both ends of the branch and use the middle section. End growth is too succulent for grafting and wood closer to the base will graft more slowly. Label the scion wood, wrap it in moist paper towels or sphagnum peat, enclose it in an airtight, plastic container and place it in the refrigerator.

Scion wood must be joined to the understock in spring when buds swell. It is critical that the two pieces are nearly the same size and that sap has begun to flow. The day before actually grafting, remove scions from the refrigerator and snip off the bottom ends. Place the clipped scions in a pail of water overnight. It is critical that the cambium layer on the scion precisely matches that of the understock.

Grafts can be placed on a root or stem and be joined by any of several methods. The union is held firmly in place using wax, tape, plastic wrap, rubber bands or plastic tubing.

The grafted area must be protected from anything that will move the scion out of alignment. Bracing some grafts is advised. Side sprouts should be removed the first summer. Sucker shoots that sprout from below the graft also should be removed. If growth of new graft is satisfactory, do not fertilise the plant during the first year. There is danger that top growth will be very brittle and it could fail to harden off before the first frost.

Bud Grafting

Bud grafting is faster, easier and less messy than other forms of grafting. Cambium layers do not need to be aligned. Bud grafting is done from early July through early August. This method uses a newly developed latent bud, taken from under a live leaf. Budwood is collected from healthy branches that grew since spring. Young trees offer good budwood because they produce a large amounts of new growth. Use buds from the middle section of the branch.

Remove leaves, but keep 1/4 inch of the leaf's petiole for grasping. The bud is cut from the branch and inserted into a T-shaped slice made in the bark of the understock. Budding should only be done when the bark slips easily away from the tree. The bud is held in place with special tape or wrap.

Micropropagation

Micropropagation is the practice of rapidly multiplying stock plant material to produce a large number of progeny plants, using modern plant tissue culture methods. Micropropagation is used to multiply novel plants, such as those that have been genetically modified or bred through conventional plant breeding methods. It is also used to provide a sufficient number of plantlets for planting from a stock plant which does not produce seeds, or does not respond well to vegetative propagation.

Each plant cell has the potential to grow into a new plant exactly like the parent. This fact coupled with technical advances, specialised equipment and sterile laboratory conditions has produced tissue culture. In tissue culture, individual or small groups of plant cells are manipulated so they each produce a new plant. A tiny piece of bud, leaf or stem can produce incredible numbers of new plants in a small space in a short time.

The advantages of tissue culture, in addition to speed and efficiency of propagation, include production of disease-free plants. New plants can be made available to the public more quickly because of tissue culture.

However, there are some problems with spontaneous mutations which naturally occur. In tissue culture, the incidence of these mutations is greatly increased.

Conditions for tissue culture are very exacting. Absolutely sterile conditions must be maintained. Temperature, light, humidity and atmosphere are strictly controlled with electronic sensors and computerised controls. Such costly equipment rules this out for most home gardeners.

Establishment

Micropropagation begins with the collection of a sterile explant(s). This small portion of plant tissue, which may be as small as a cell, is placed on a growth medium, typically a medium containing sucrose as an energy source and one or more plant growth regulators (plant hormones).Usually the medium is thickened with Agar to create a gel which supports the explant during growth.

The plant tissue should now begin to grow and differentiate into new tissues. For example, media containing cytokinin are used to create branched shoots from plant buds.

Multiplication

Following the successful growth of plant tissue, the establishment stage may be repeated, by taking tissue samples from the plantlets produced in the first stage.

Through repeated cycles of this process, a single cell sample may be magnified to hundreds or thousands of plants.

Pretransplant

This stage involves treating the plantlets/shoots produced to encourage root growth and "hardening". It is performed in vitro, or in a sterile "test tube" environment.

Root growth does not always occur in the earlier stages in plant cell culture, and is of course a requirement for successful plant growth after the micropropagation procedure. It is performed in vitro by transferring the plantlets to a growth medium containing auxin(s).

"Hardening" refers to the preparation of the plants for a natural growth environment. Until this stage, the plantlets have been grown in "ideal" conditions, designed to encourage rapid growth. Due to lack of necessity, the plants are likely to be highly susceptible to disease and will be inefficient in their use of water and energy.

Hardening typically involves slowly weaning the plantlets from a high-humidity, low light, warm environment to what would be considered a normal growth environment for the species in question.

This stage (pretransplant) is not always performed, instead being incorporated into the last stage by encouraging root growth and hardening ex vitro, or in nonsterile plant media.

Transfer from Culture

In the final stage of plant micropropagation, the plantlets are removed from the plant media and transferred to soil or (more commonly) potting compost for continued growth by conventional methods. This stage is often combined with the "Pretransplant" stage.

Advantages of Micropropagation

Micropropagation has a number of advantages over traditional plant propagation techniques:

- Micropropagation produces disease-free plants
- Micropropagation produces rooted plantlets ready for growth, rather than seeds or cuttings
- It has an extraordinarily high fecundity, producing thousands of propagules in the same time it would take a conventional technique to produce tens or hundreds
- It is the only viable method of regenerating genetically modified cells or cells after protoplast fusion
- It is a good way of multiplying plants which produce seeds in uneconomical amounts (if at all)
- Micropropagation often produces more robust plants, leading to accelerated growth compared to similar plants produced by conventional methods

Disadvantages of Micropropagation

Micropropagation may appear to be the perfect means of multiplying plants, but it has associated problems:

- It is very expensive, and can have a labour cost of more than 70%.
- An infected plant sample can produce infected progeny. This is uncommon, as stocks are usually carefully vetted to prevent this.

The greatest limitation is the cost. Most plants will naturally produce seeds, which are normally disease free and will readily grow under good conditions. The number of seeds varies, but is normally acceptable for multiplication and is free. For this reason, many plant breeders will never resort to micropropagation because of the prohibitive cost. Mechanisation of the process would eliminate most of the labour cost associated, but this has proven difficult so far despite active attempts to develop this technology.

7

Plant Growth Regulators in Floriculture

Plant hormones, or plant growth regulators, are chemicals produced by plants that alter growth patterns and/or maintenance of the plant. They can be found in many cells and tissues, although plant hormones seem to be concentrated in meristems and buds (which are dormant shoot meristems). Growth regulators control cell activities by sending chemical signals or messages to cells to do something or to not do something, including activating the genes that code for specific enzymes. Plant hormones inhibit as well as promote cellular activities.

In contrast to animal hormones, which generally have very specific effects, the hormones identified in plants most often regulate division, elongation and differentiation of cells. Most hormones have multiple effects in plants. Plant hormones work in very small concentrations. In most cases, the effect plant hormones have on the plant depends on the location of and concentration of the hormone relative to other hormones in the specific tissues. Plant hormones often work in conjunction with each other, and have overlapping effects. They also work with environmental stimuli, as we shall discuss.

Classes of Plant Hormones

There are several classes of plant hormones. Let us look at each of these briefly.

Auxins

The concept of chemical messengers in plants was proposed by Charles Darwin and his son, Francis, in 1880, who spent time looking at the phenomenon of phototropism in wheat seedlings. Plant shoots are positively phototropic. When a seedling is illuminated from the side, the shoot will bend towards the light. This directional growth makes sense, since plants need light for photosynthesis.

However, the Darwins found that if the coleoptile of the wheat seedling was removed, the plant no longer curved towards light. They did a number of experiments and determined that a chemical located in the coleoptile traveled to the region of elongation and effected a differential elongation of cells furthest from the light sources. In 1913 Peter Boysen-Jensen affirmed that the signal for bending was mobile. When he placed a mica slice between the coleoptile tip and the rest of the shoot, no bending occurred. The mobile chemical was subsequently studied and named auxin by Frits Went in 1926.

Chemically auxin is indoleacetic acid (IAA), which is synthesised from indole or tryptophan. There are a number of synthetic "auxins" too. Auxins promote growth in molar concentrations of 10-3 to 10-8. Auxins are synthesised in shoot apical meristems, young leaves, seeds and fruits. Auxin always moves down the stem parenchyma cells towards roots by polar (charged) transport (auxin becomes negatively charged) using proton pumps, an ATP requiring process.

Auxin enters cells as IAAH passively, or as IAA- via active cotransport. IAAH dissociates within the cytosol and special auxin transport proteins in the basal end of the cell are needed to carry auxin through the plasma membrane to the top of the adjacent cell. Auxin destined for root tissue, however, moves through phloem sieve tubes.

Auxin promotes elongation and cell enlargement. For cell elongation, proton pumps increase the H+ concentration in cell walls, which stimulates expansins, proteins that disrupt hydrogen bonds and break cross linkages in cellulose. This facilitates wall expansion when cells take in more water.

Auxins are involved in tropic responses. Auxins migrate away from light, which accounts for the uneven elongation of cells on the shaded side of a plant unevenly exposed to light, as studied by Darwin.

Auxin stimulates the production of secondary growth by simulating cambium cells to divide and secondary xylem to differentiate. Wound tissue repair is initiated by auxin when portions of vascular bundles are damaged.

Auxin produced in apical buds tends to inhibit the activation of buds lower on the stems. This is known as apical dominance. This effect lessens with distance from the shoot tip. Cytokinins (another group of plant hormones) counter the apical dominance effect of auxins and promote lateral bud development.

Loss of auxin initiates leaf abscission. Fruit development requires auxin produced by the developing seed. Auxin pastes applied to developing ovaries can promote parthenocarpy (fruit development in the absence of viable seeds). Parthenocarpy is also induced by abscisic acid.

Auxins are toxic in large concentrations. They affect mostly dicots but not monocots. Monocots seem to be able to rapidly degrade the synthetic auxins. Synthetic auxin herbicides include 2-4-D and 2-4-5-T (Agent orange of the Vietnam era contains synthetic auxin).

Cytokinins

Cytokinins are a group of phenyl urea derivatives of adenine, one of the molecules in DNA. The first cytokinin was chemically isolated in 1913 and cytokinins were studied using coconut endosperm for a number of years starting in the 1940's. This isolate was shown to be a potent growth promoter and was used in tissue culture and embryo development studies.

Folke Skoog and Carlos Miller spent several years attempting to isolate this growth substance. They finally succeeded in isolating a breakdown product of DNA that promoted cell division. They named the substance kinetin and the related group of growth regulators, cytokinins, because they were involved in cell division. However natural cytokinins were not chemically identified until 1963, when zeatin was identified in corn.

Cytokinins are found in actively dividing tissues of seeds, fruits, leaves and root tips, and wound tissue sites. Studies indicate that root tips are most likely the location for cytokinin production and cytokinins are transported through xylem to the rest of the plant. However, localised cytokinins are needed to release buds from dormancy.

Cytokinins promote cytokinesis. They are found primarily in root meristems, embryos and fruits, and migrate from roots to the shoot systems of plants in xylem tissue. The effects of cytokinins are often studied in tissue culture. Normally, cytokinins work in conjunction with auxins, which promote elongation and cell expansion. When auxin is added to a tissue culture medium, cells in the absence of cytokinins enlarge, but division does not take place. When kinetin and IAA are both added to a tissue culture medium, cells divide rapidly producing many small cells. The ratio of cytokinins to auxin controls what tissues differentiate in the tissue culture.

Cytokinins can promote axillary bud growth by overriding the apparent inhibiting effect of auxin. This is one of the ways in which plants balance root and shoot growth. Shoot tip auxins inhibit lateral bud activation. Cytokinins produced in root meristems travel upward in xylem and if in higher concentration counter the inhibition of auxins to activate lateral buds.

Cytokinins also retard leaf senescence, probably by stimulating RNA and protein synthesis and delaying degradation of chlorophyll. Some pathogens that promote tumor formation, called galls in plants, take advantage of cytokinin function. The bacteria that form these tumors contain genes for the synthesis of cytokinins so promote rapid undifferentiated cell growth in the infected part of the plant. Agrobacterium tumefaciens, a common vector in plant biotechnology contains such genes.

Gibberellins

Ewiti Kurosawa is credited with the discovery of gibberellins when he determined that a fungus was responsible for abnormal rice seedling growth, called the "foolish seedling" disease. The fungus secreted a chemical that caused the rice plants to grew abnormally long, and then collapse from weakness. The fungus was Gibberella fujikuroi, hence the hormone name.

Many seeds contain a variety of different gibberellins. Over 100 different gibberellins (organic acids synthesised from mevalonic acid) are known. Gibberellins are produced in roots and younger leaves, but have their highest concentration in seeds. Most effects of gibberellins are shown only in concert with auxins.

Gibberellins work with auxins to promote rapid elongation and division of stem tissue. Gibberellins determine microtubule alignment in the preprophase band that determines the plane of cellulose expansion.

The effects of gibberellins on elongation are seen in:

— Bolting of biennials, to produce flowers during the first growing season

— Reversal of genetic dwarfism

Gibberellins are important to breaking dormancy after imbibition of water by the seed coat. Gibberellins signal germination activities. In particular, application of gibberellins to seeds will counter the normal environmental cues, such as exposure to low temperatures. They stimulate RNA to promote synthesis of enzymes that convert stored nutrients (starches) to sugars needed for rapid cell respiration during germination. Gibberellins can, as can auxins, promote parthenocarpy in many plants, which has commercial applications, although the most significant use in the grape industry, where grapes grow larger, and with longer internodes.

Abscisic Acid

Abscisic acid (ABA) is a hormone that functions by inhibiting growth activities in times of environmental stress rather than by promoting growth. It often serves as an antagonist to the other growth promoting hormones in plants. Abscisic acid, which is also synthesised from mevalonic acid, got its name from the erroneous belief that it promoted the formation of abscission layers in leaves and fruits. It does not, although leaf abscission accompanies dormancy in many plants.

ABA promotes seed dormancy activities. ABA levels are high when seeds mature, promoting lowered metabolism and synthesis of proteins needed to withstand the dehydration associated with dormancy. Seeds germinate when ABA is degraded by some environmental action. Desert seeds must have the ABA washed out of the seed coat, as do plants of marshy areas; temperate area plants have ABA degraded by light-stimulated enzymes. In other cases breaking dormancy is relative to the ratio of ABA (which keeps seeds in dormancy) and gibberellins (which promote germination).

Low levels of ABA in maturing seeds promote premature germination. Similar to seed dormancy, ABA promotes winter bud scale formation on woody plants in preparation for winter dormancy. ABA derivatives, called dormins, are used in commercial nurseries to keep materials to be shipped in dormant conditions. The dormancy can be reversed with gibberellins.

ABA is also referred to as the stress activity hormone. For example, ABA promotes stomata closure during leaf water deficit conditions by activating K+ ion transport out guard cells. This involves signal transduction pathways with Calcium secondary messengers. ABA in this case originates in roots, and detects low water level in root tissues. ABA moves upward into leaves and activates stomatal closure.

Ethylene

Ethylene is the sole plant growth regulator known that is a gas. Ethylene is produced in many plant organs, though its effects are most studied in fruits. Ethylene is the signal molecule for signal transduction pathways that activate a number of tissue growth and development activities throughout the plant, including fruit maturation, leaf abscission and senescence.

Ethylene is synthesised from the amino acid methionine. High auxin concentrations promote an intermediate step in this pathway that activates an enzyme in the tonoplast (vacuole membrane) to convert the intermediate into ethylene. Toxic substances such as air pollutants also trigger the intermediate step. Ethylene is also produced in wounded or bruised tissues. The benefit of this is unknown; it may be an artifact of the wounding response.

The benefits of ethylene as a fruit ripener were known for centuries prior to it being identified as a plant product in the early 1900's. Chinese gardeners knew centuries ago that fruits ripened better in rooms with burning incense. Citrus growers used kerosene stoves in the rooms in which they ripened their fruit. During the era of gas lamps, leaking lamps along city streets often promoted leaf abscission. Today, grocer warehouses have ethylene rooms that are used for ripening most of our produce, which is shipped unripe. Immature fruits are firmer and less subject to damage. Thanks to ethylene rooms, we have "mature green tomatoes".

Ethylene activates ripening in fruits by signaling chemical reactions that degrade the pectins of the middle lamella, softening fruit, and promoting the conversion of stored starches and/or oils into sugars that attract seed dispersers. Some fruit ripening involves a significant increase in the rate of cell respiration with concurrent high O2 uptake. This function of ethylene (fruit ripening) has extensive agricultural impact. Biotechnology has been used to alter sensitivity to ethylene in some crops to facilitate harvest and shipping of fruits. Using gene technology, tomatoes lacking ethylene receptor genes have been grown and their fruits have a shelf life of 100 days!

Ethylene promotes female flower production in some members of the Cucurbitaceae, whereas high gibberellins may promote formation of male flowers. Ethylene functions to help germinating seeds overcome mechanical stress. A shoot tip that encounters an immovable object will grow around the object by changing its growth direction. This occurs through differential elongation of cell walls. When the shoot tip cannot penetrate the mechanical obstacle, ethylene is synthesised, which slows cell wall expansion. The walls thicken (more resistant to the pressure) and the stem curves to bypass the obstacle via horizontal growth. (Essentially the shoot is growing around the obstacle.) The stem tip pushes upward periodically to test for the obstacle. If the obstacle is encountered, ethylene initiates another cycle of stem thickening and horizontal growth. If no obstacle is encountered, the shoot resumes negative geotropic growth. This growth response is known as the triple response (decrease in elongation, stem thickening, and change in orientation of shoot growth to horizontal).

Ethylene is responsible for initiating the programmed death (apoptosis) of sclerenchyma and xylem vessels and tracheids. Death involves intense cellular activity to degrade and salvage materials of the cytoplasm.

Ethylene is the direct cause of leaf and fruit abscission. Ethylene promotes the degradation of the cell walls in the abscission zone cells. The declining levels of auxin in leaves signal the production of ethylene in the abscission zone cells as summer ends.

The auxin/ethylene interaction for abscission is also used commercially. Fruit growers may spray auxins on fruits to prevent the fruits from falling to the ground prior to harvest. Concentration is important; very high auxin level promotes ethylene production. High concentrations of CO_2 inhibit ethylene production so CO_2 can be used to prevent fruit ripening in grocer warehouses.

Brassinol Ides

Brassinolides is a plant steroid discovered in pollen of members of the mustard family, and studied in Arabadopsis. Chemically the brassinosteroids are similar to animal steroid hormones, and have similar genes that code for the steroid synthesis pathways. The functions of brassinosteorids overlap those of other plant hormones, and may be additive. Brassinosteoroids activate signal transduction pathways that promote cell elongation and cell division. Brassinosteroids also promote differentiation of xylem tissue, and perhaps other tissues, too. Brassinosteroids can also retard leaf abscission. Absence of brassinolides results in dwarf plants. It is difficult to study the brassinosteroids because their effects overlap those of auxin and gibberellins.

Salicylic Acid

Salicylic acid is known to activate defense genes for resistance against pathogen invaders, known as the hypersensitive response. Salicylic acid, a phenolic extract from willow bark, was long used as an analgesic. It is now prepared commercially and is the active ingredient of aspirin.

Oligosaccharins (Oligosaccharides)

Oligosaccharins are short chain sugars in cell walls that

may have a role in defense against pathogens. When a pathogen "chomps" on the oligosaccharide, the oligosaccharide activates a signal transduction pathway leading to plant defense responses. They may also help regulate growth, differentiation and flower development all by activating signal transduction pathways.

Systemin

Systemin is a small peptide found in wound tissue. It may stimulate defense activities as the signal molecule (ligand) that activates the signal transduction pathways that include the jasmonates.

Jasmonates

Jasmonates are a group of fatty acid derivatives. They appear to have a role in seed germination, root growth, and the storage of protein (especially in seeds). Jasmonates synthesised in response to signal molecules produced in wound areas are involved in the signal transduction pathways that result in secondary metabolites (protease inhibitors) that poison the predator, or synthesis of volatile molecules that attract the predator's predators.

Plant Growth Regulators and Genetic Engineering

The new technologies of genetic engineering can result in more environmentally friendly plant pest-protection, foods with enhanced nutrition, more accurate and sensitive diagnostics, foods with improved processing and marketing characteristics, better and more efficient medical delivery, new methods for removing contaminants from soils and waters, and creation of products that are presently being made from nonrenewable resources. Many of these changes might be considered under the broad definition of a "plant growth regulator".

While products of this technology once were confined to the research laboratory, this is no longer the case. These and other new crops represent a substantial percentage of actual production acreage in the U.S. In 1998 50% of the cotton acreage, 30% of the soybean and 20% of the maize acreage in the U.S. were genetically engineered. In the U.S. some 60 million acres of G.E. crops were planted in 1998 and the projections for 1999 are that this number will increase, with a wider variety of products available to farmers and present in the marketplace.

Pest Control

Herbicide tolerance

A major agricultural advance occurred with the identification of selective herbicides that can be used to eliminate weeds without affecting the crop plant. But achieving complete selectivity has been problematic for many crops. An alternative to finding new selective herbicides is to genetically engineer the crop to resist existing herbicides. The engineered mechanisms can include modifications in the site of action of the herbicide, in its detoxification or in the uptake and translocation of the herbicide to its site of action. Such crops represent the first and most frequent application of biotechnology to agriculture. Of the total of 5900 permits and notifications issued through the end of 1998 by the USDA Animal and Plant Health Inspection Service (APHIS), 27% or 1569 permits were for herbicide-tolerant crops. Based on data from APHIS, work is currently being done on six herbicides: bromoxynil, glyphosate, imadazolinones, phosphinothricin

In many cases, the herbicide tolerance is based on a modified site of action through the use of mutant or altered forms of the target protein. In these instances, the altered enzyme must be properly compartmentalised (chloroplast, cytoplasm, nucleus) and it must be assembled in a metabolically active form at its target location. This is the case with glyphosate where the herbicide blocks a pathway that leads to the synthesis of several amino acids; blockage of this pathway leads to starvation for amino acids. Genetic engineering strategies have focused mainly on using mutant target enzymes that are resistant to glyphosate. A similar type of strategy is used with another group of herbicides, the imadazolinones and sulfonylureas, which also inhibit a key step in the biosynthesis of a different set of amino acids.

Other approaches to engineering herbicide resistance involve the introduction of genes for enzymes that degrade and detoxify the herbicides. One example in this category is bromoxynil in which resistance was engineered through the introduction of a bacterial gene that codes for an enzyme that converts bromoxynil to an inactive metabolite. Another example is glufosinate. Resistance to this herbicide can be engineered using a bacterial gene that acetylates and inactivates the herbicide. The last example is 2,4-D, a plant hormone that causes arrested development. Again, genetic engineering strategies employ a bacterial gene that encodes an enzyme that degrades 2,4-D. Cotton has been engineered to be resistant to this herbicide in order to prevent crop loss due to spray drift.

Although progress in engineering herbicide-resistant crops has been rapid, there are some problems with this approach. Technical problems include finding an effective strategy for engineering resistance and identifying the target site of the herbicide or a degradative or inactivating factor. In addition the resistance factor must be engineered to be expressed in the proper place and in sufficient levels.

The herbicides for which tolerance is being engineered are generally low-use rate, low-toxicity, rapid-turnover herbicides. Despite this, there are some environmental issues which, although not unique to genetically engineered varieties, must be taken into consideration.

— *Generation of herbicide-tolerant weeds due to monoherbicide application.* Frequent and long-term application of a single herbicide becomes more likely because of the use of specific herbicide-tolerant varieties. The tendency of the industry to engineer a large number of crop species with tolerance to a single herbicide will create pressure for this to occur

— *Transfer of resistance genes to related weeds.* Many crops in the U.S. in which tolerance is being developed do not have weedy relatives here; exceptions are sugarbeet, certain cucurbits, radish, canola, oats, and carrots. As seed is moved internationally, however, the possibility of outcrossing with wild relatives in centers of diversity will increase; control measures should be instituted. In general, however, unless pollen travels over long distances to pollinate weeds located far away, this problem will be minimal.

— *Encroachment of weeds not affected by the herbicide, i.e. weed shift.* Development of elite germplasm resistant to a single herbicide, without comparable varieties resistant to other herbicides with dissimilar modes of action, could lead to an exacerbation of the weed shift problem.

Insect resistance

A now standard method for engineering insect resistance in crop species is the use of the toxin genes from Bacillus thuringiensis; this is the basis for Bollgard cotton and the commercially available maize that is European corn borer-resistant. These toxins are activated by cleavage in the midgut of insects; however, making engineered toxins that are effective against certain insects has not been straightforward. For example for a certain insect pest of alfalfa, Spodoptera armyworm, the situation was not straightforward. It required some extensive in vitro testing of synthetic toxins. In the end a specific synthetic delta-endotoxin gene was identified, which when expressed in optimum amounts in alfalfa, shows evidence of protection against the Egyptian cotton leafworm (Spodoptera littoralis) and the beet armyworm (Spodoptera exigua). In other cases new types of Bt's are being discovered and also some insecticidal compounds from other microbial species.

Viral resistance

Infection with viruses is a problem with many crop species; however the sources of natural resistance are limited and chemical protection is not effective in many instances. A standard method for engineering viral resistance is the use of the coat protein gene. This approach has been used successfully for many viruses in many different plant species and the Freedom II squash was released in the marketplace last year. In the coming year genetically engineered papaya will enter the marketplace. This crop results from efforts to protect it against papaya ringspot virus, which had devastated certain production areas in Hawaii. Because there are limitations to the coat protein approach in terms of uniform applicability and broad protection to related viruses, it is necessary to develop alternative strategies that complement or replace the coat protein strategy. These have included the use of viral polymerase and movement protein genes; such strategies require an in-depth understanding of the architecture of the

Fungal and bacterial resistance

Methods for protecting against fungal pathogens involve the identification of natural plant protection compounds that, when expressed in higher amounts in appropriate tissues of the host plant, can lessen fungal damage. These are sometimes compounds deriving from secondary metabolic pathways. Work on compounds, such as these, can lead to approaches where increased or broader-spectrum disease resistance can result from modifying expression levels or timing of unmodified secondary metabolite or modifying the phytoalexins themselves to create new phytoalexins with novel activities.

In addition natural defense genes have been isolated for resistance to various pathogens in different plant species Of the first three such genes that were isolated, it was discovered that they share common domains. Understanding these domains and their function, it will be possible to devise plant protection strategies using these genes that provides broad spectrum resistance against a variety of pathogens.

Nematode resistance

A variety of approaches have been utilised to protect against damage from nematodes. Regulatory regions have been identified which are uniquely engaged when the nematode burrows into the root; simple physical damage of the root is not sufficient to turn on the promoter. This regulatory region can then be hooked to nematicidal proteins, which can then be delivered to the nematode at its precise site of attack. Such nematicidal proteins include several different proteinase inhibitors and lectins, which act in some cases synergitically to interfere with the sexual development of the nematode.

In addition, recently a natural resistance gene was isolated from tomato and found to provide protection in engineered plants to not only nematodes but also aphids, which have an entirely different mode of attack and damage pathway. It was found, after comparison of the gene to other natural resistance genes from other species, that the nematode resistance gene shared common domains with these other genes, again pointing to the potential universality of these mechanisms and opening the door for creating some engineered proteins with widespread efficacy.

Harvest and Post-harvest Characteristics

Another arena in which we have seen product enter the marketplace and in which much research is being conducted relates to harvest and post-harvest characteristics. Two different approaches to delayed ripening have been commercialised. The approach of the Flavr Savr tomato utilised a technique allowing growers and producers extra time to pick and deliver tomatoes to the marketplace by interfering with polygalacturonase synthesis, an enzyme involved in pectin deposition. In this case the ripening process was delayed by several days but was not halted. In another approach to interfering with the ripening process, researchers stopped the ripening process completely by interfering with the production of ethylene, which is essential to ripening. By this process it is possible to leave a tomato on the vine for up to 100 days past the time when it would ordinarily be ripe and, once it is picked, to ripen the tomato on demand by exposure to ethylene.

Another post-harvest characteristic, which is the focus of genetic engineering technologies is the achievement of more uniform ripening. This was first accomplished with the cherry tomato, but the basic approach is also being attempted with processing tomatoes in order to achieve less waste and aid harvesting. This approach involves the manipulation of genes involved in slowing the degradation of ethylene.

Agronomic Traits

Nitrogen fixation and utilisation. Fundamental work on the process of nitrogen fixation has been pursued for decades. With the advent of molecular techniques, pace of that research has increased and has broadened to include fundamental studies on the nature of the plant-microbe interaction necessary for nodule formation and nitrogen fixation.

Work on the bacterium itself has resulted in engineered Rhizobium strains that have increased rates of nitrogen fixation in limited field trials, but those strains have not yet been released for widespread use. Studies focused on understanding and manipulating the host range of nitrogen-fixing bacteria are also underway with the expectation of being able to extend the range of crop species that can have productive commensal relationships with bacterial nitrogen fixers.

In addition to looking at improving the efficiency of nitrogen fixation, fundamental work aimed at understanding nitrogen fertiliser use efficiency is also underway. This work hopes to lead to engineered plants, e.g. wheat and corn, that more efficiently use applied fertilisers.

Male Sterility

The development of hybrid varieties that result in increased yields of crop species has been a tedious process, requiring in many cases the manual detasseling of the male parent to prevent self-fertilisation. Large scale production of hybrid seed, however, required more efficient mechanisms to suppress pollen production. Classical geneticists have identified naturally occurring male-sterile lines of some crops that were used for commercial seed production. Such strategies also required the identification of "restorer lines" that could restore male fertility by counteracting the genes that cause male-sterility. This is more easily accomplished with some crops than with others.

With the advent of genetic engineering, it became possible to develop strategies to create male sterile plants. These strategies made use of the fact that one could identify regulatory sequences that would lead to very specific tissue expression. In one strategy a promoter was identified that only drove expression in the inner layer of the pollen sac, the tapetum. Linked to this promoter was a bacterial enzyme, which degrades RNA, thereby interrupting the development pollen grains. A restorer of fertility was also identified; this protein is a chemical coupler of the RNA-degrading enzyme; the tight coupling of these two proteins inactivates the RNA-degrading enzyme, thus restoring fertility. This approach was used in canola first and is now being attempted in a variety of different crop and vegetable species.

Terminator Technology

The fundamental idea with this technology is to prevent a plant from reproducing itself by interfering with seed development, thus insuring that the user of the seed cannot replant a field without repurchasing the seed, thereby protecting the company's investment in developing the variety. The system works through the interplay of three introduced genes. One of these genes leads to the production of a toxin that kills the seed in the final stage of development. This gene is initially interrupted by a stretch of DNA between the toxin gene itself and the regulatory region or promoter that is responsible for its synthesis.

This gene configuration can then be transplanted or crossed with another genetically engineered plant that produces a scissor-like enzyme, capable of removing the segment in the toxin gene, thereby activating it and causing the seed to die. The third component of the system is a repressor gene that prevents the synthesis of the scissor-like enzyme. If all three components are present, the terminator technology is dormant and the seed is fertile.

Value-added Products

There is another area that does not involve plant growth regulators but is being intensely pursued by both the public and private sectors. This relates to the improvement of existing plant products, ones that already exist but for which new or improved functions can be identified that increase the value of the product. An example would be specialty oils from rape or soybean, which are designed to have improved heat stability. Another example would be the genetic engineering of sugarbeet to produce fructans, a sweet, no-calorie sugar substitute.

An intriguing application is in the area of functional foods or neutriceuticals. This line of foods is based on the fact that we can now identify in many cases natural products in foods that can either cause adverse human condition or prevent disease. Examples of those products focused on factors in foods that cause disease or undesirable pharmacological effects include ameliorating the allergenic potential of foods, such as wheat, milk and peanut or removing caffeine from coffee. In the category of natural products that prevent disease, there are many examples.

Transgenic potatoes are being created which produce high quantities of GAD, an enzyme believed to be involved in the prevention of type II juvenile diabetes. Researchers have identified an element in grapes that increases blood flow in blood vessels; grapes could be engineered to increase the level of such an element. Several groups are coming up with margarine and oils, based on natural plant sterols and stanols, that are not only "low in choles.

Managing PGR Use in Floriculture

PGRs are chemicals used on a wide range of floricultural crops. Products currently on the market are labeled for control of plant height, for stimulation of lateral branching (Florel), for substituting for a cold storage requirement (GibGro 4LS), or for promoting flower initiation or earlier flowering (B-Nine, Cycocel, Florel, and GibGro 4LS). By far, the majority of plant growth regulators employed in floriculture are chemical growth retardants used for control of plant height. However, the application concepts and techniques are given below for all PGR's labeled for use on floricultural crops. While digesting the following text, keep in mind that chemical growth retardants (CGR's) are a class of plant growth regulators (PGR's); not all PGR's are CGR's.

Before applying a PGR, a grower should consider the reason for using the PGR. Recall that PGR's are applied to plants to regulate plant development and to stimulate a desired growth response. For example, chemical growth retardants are used to retard growth, resulting in shorter plants. A closer look at how CGR's work may help emphasise the importance of correct choice and use of PGR's.

Most of the available growth retardants are anti-gibberellins; i.e., they inhibit the synthesis of gibberellins such as gibberellic acid (GA_3) within the plant. Gibberellins stimulate cellular elongation, so without them, cells do not elongate as much, and plants do not grow as tall. Ethephon is not an anti-gibberellin; ethephon releases ethylene, which reduces elongation in some crops.

Since CGR's, as do all PGR's, affect a specific process in the plant, it is essential they be applied in a manner that assures the most efficient response. Both monetary and environmental costs are too great to apply PGR's carelessly. This conscientious effort to use the minimum amount of chemical as effectively as possible is the basis for Best Management Practices. Best Management Practices for using PGR's can be divided into different categories: timing, target tissue, dosage, application technique, and environmental conditions.

Timing

Timing of PGR applications must be matched to the proper plant stage of development to achieve the desired goal. Usually the labels on the products will give good descriptive stages of plant development to assist with correct application timing. For example, the GibGro label also defines when to make applications. If GibGro 4LS is being used to partially substitute for cold storage of azaleas, plants should be at Stage 5 of floral development (i.e., style elongated and open) when treatment is initiated. Applying prior to this stage could result in abnormal flower development.

For most plants, apply CGR's such as A-Rest, B-Nine, Bonzi, Cycocel, Florel, and Sumagic after the plant has developed sufficient foliage (photosynthetic area, leaf area) to prevent excessive stunting of the plant's development. For example, a Bonzi spray on pansy plugs should not be made until three true leaves are present on the plants. Earlier applications may stop plant growth completely rather than merely reduce elongation. Apply CGR's just prior to rapid shoot elongation; e.g. after pinching and newly developing shoots are visible, but before shoot elongation has occurred.

Remember that CGR's are chemical growth retardants not chemical plant reducers—they cannot shrink plant growth already present. Make your final application before the stage when floral size will be reduced. If growth retardants are applied too late, the size of the flowers can be reduced and floral development can be slowed. A good example would be bract size reduction and delay of bract colour in poinsettia due to late application of A-Rest, B-Nine + Cycocel, Bonzi, Cycocel, or Sumagic.

Timing of a PGR application should be based on a physiological stage of plant development such as the number of leaves present, the length of shoots, or plant diameter, not chronological age such as 3 weeks after pinching. Recommendations given in terms of chronological age are merely guidelines that have been correlated to physiological stages and should only be used as rough estimates as to when to apply plant growth regulators. Also remember to read the label when deciding on timing for a PGR application. Many labels suggest the proper plant stage of development for chemical application.

Target

The target tissue or plant part to receive the chemical depends on the chemical and the plant species.

— *Foliage and stems:* Chemicals can be sprayed on, or shoots can be dipped into a chemical solution at time of transplanting. If dipping shoots into a solution, use a consistent soaking time, such as 10 seconds. For sprays, remember that areas are sprayed not individual plants. Imagine yourself painting a wall with a roller. You want uniform coverage over the entire area—not heavy clumps. This is the only way to assure even coverage and consistent results. With many PGR's such as A-Rest, Bonzi, and Sumagic, spraying different volumes of the same concentration spray over a given area gives different levels of control. If growers "spray to runoff" or "spray to glistening", every grower will apply a different amount, and there is no way of predicting the results.

  Proper coverage is essential for consistent results. For example, Bonzi and Sumagic are not very effective if only applied to plant leaves. Transport of Bonzi and Sumagic to the growing point in the plant is most effective if it is applied to the stem or if it is absorbed through the roots. If an insufficient application volume is used and stems are not wetted, poor height control will be the result. On the other hand, since root uptake is very effective, an over application of Bonzi or Sumagic using too large a volume will result in the excess spray dripping onto the substrate leading to root uptake of the chemical and causing too great a reduction in growth.

— *Substrate:* PGR's can be applied to the substrate (growth media or soil), if the root system is the target tissue. This is the same as a soil drench; drenches use larger volumes of solution per plant or pot, but usually at lower concentrations than a spray or dip. Drenches can take more time to apply than sprays, and require exact metering of volume delivered per pot for consistent control of elongation.

— *Roots:* PGR's can be applied directly to the roots or the underground portion of the plant such as bulbs, corms, or tubers, prior to planting. An example is a Bonzi soak, which is labeled as a preplant treatment for freesias.

Dosage

Read the label; do not guess on dosage. Keep in mind that a dosage is the product of concentration of solution applied and volume of solution applied per area. If either are incorrect, results could be unpredictable and nonrepeatable. Take care to always apply the correct dosage to all plants treated.

Application Technique

As with the target tissue, the method of delivering a PGR depends on both the chemical used and the plant species to be treated.

— *Dips:* With some plants it is possible to dip the plant shoot, or underground portion (bulb, corm, tuber) into a growth retardant solution prior to potting. This method is labeled and is very effective for applying B-Nine to the shoots of rooted chrysanthemums just prior to potting. This method is fairly accurate, if each plant remains in the solution for the same amount of time, and if each plant has approximately the same size shoot or bulb. Unfortunately, this method is not feasible with many crops, and dipping plants in a common solution could result in the spread of disease organisms.

— *Drenches:* Applying a growth retardant in a drench form is fairly easy. Measure out a known amount of chemical, add it to a known volume of water, and apply a known volume of the drench to each pot or plant. As pot size increases, usually the volume of drench recommended also increases. If using a drench, make sure the volume and concentration of the solution are correct for the size pot to be treated. There are "drenching machines" available on the market that emit a set volume of drench then turn off for a short period of time prior to the next dose to allow applicators to place the nozzle into the next pot to receive the drench. Do not apply any plant growth regulator through an irrigation system, unless the label lists chemigation as a legal application technique. Currently, it is legal to apply A-Rest through drip, overhead boom, sprinkler, and flood systems. Consult the label for specific application recommendations.

— *Sprays:* A spray application can be more difficult to apply evenly than a drench, but with attention to detail, is the method of choice. Some chemical labels recommend to "spray to runoff"; that is, spray each plant until spray visibly just begins to drip off of the foliage. Depending on the size of the plant, the sprayer's objectivity, and other factors, varying amounts of chemical will be applied to each plant. It is much safer and more accurate to base spray application on areas, not plants. Apply a known volume of spray evenly to a known area (square footage), regardless of how many plants are in that known area. The general recommendation for sprays is to apply $^1/_2$ gallon per 100 ft^2 of cropping area. This volume is sufficient to comfortably walk 25 feet while spraying a 4 foot-wide bench, thus the basis for the recommendation. If the area is sprayed evenly, it assures that each pot will receive the same amount of spray, regardless of how many containers are in the area.

To help with uniform distribution, the sprayer should be equipped with a pressure gauge and pressure regulator to assure uniform output. Unless the sprayer is emitting a constant volume, spray will not be applied evenly over plants.

Spray droplet size can affect spray distribution on the plants. In general, the smaller the droplet size, the more even the coverage and the greater the effectiveness (efficacy) of the chemical. Therefore it is possible that two growers applying the same treatment but with different spray nozzles could achieve different degrees of effectiveness. For consistency, try to use the same nozzle type for all PGR applications.

Spray adjuvants, compounds added to assist the action of the active ingredient, such as a wetting agent can affect PGR efficacy. Some PGR products such as A-Rest, Bonzi, and Sumagic advise against adding a wetting agent. B-Nine already contains a wetting agent and may not require the use of an additional wetting agent. If a wetting agent is needed to allow a spray material to more thoroughly cover plant surfaces, such as with a Cycocel spray on poinsettias, use as little as possible to avoid phytotoxic side effects. Start off using a very low concentration and by experimenting with water on test plants, slowly increase the wetting agent concentration. Add only enough wetting agent to prevent droplets from beading up on the leaves; a higher concentration is unnecessary and could cause leaf burning.

Environmental conditions: The efficacy of a PGR can be affected by:

1. the environment at time of application,
2. the status of the plant at application, and
3. post-application treatment of the plant.

A good example of an environmental factor affecting PGR efficacy is the effect substrate components can have on CGR's. A bark-containing substrate will reduce the efficacy of A-Rest and triazole growth retardants like Bonzi and Sumagic, when applied as a drench. Therefore, drenching may not be the application method of choice for these PGR's if using a bark-based substrate; or growers should account for the presence of bark when deciding on the concentration of PGR to use.

Another environmental factor of concern is the time of day selected for applying PGR's. Research with foliar sprays of nutrients has shown that the time of day chosen for an application can affect a plant's ability to absorb a chemical. Morning applications, when evaporation rates are low, are more desirable than later in the day. The longer the solution wets the tissue, the greater the chance for chemical uptake. If possible, make spray applications on a cloudy day to allow the chemical solution even more time to be absorbed prior to evaporation of the water from the surface of leaves and stems.

Plant water status can affect chemical efficacy. Plants should not be wilted or stressed at application. A turgid plant is more able to absorb and translocate a PGR than a wilted plant.

With respect to post-application handling of treated plants, B-Nine has been shown to require a long period of time (up to 4 hours) for complete absorption after a spray application. If the plants are watered (wetting the foliage) too soon after a B-Nine application, the unabsorbed chemical will be washed off and the efficacy of the B-Nine will be reduced. This is not true for rapidly absorbed PGR's such as A-Rest, Bonzi, and Sumagic, and plants either drenched or sprayed with these chemicals can be watered as soon as 1 hour after application without a reduction in chemical efficacy. The effects of post-application watering on Florel and GibGro efficacy are not known.

Problems Due to Plant Growth Regulators

Plant growth regulators are excellent tools to help control undesirable stretch of floriculture crops. When used at the correct stage of development and rate, they provide excellent height control. Misapplications can lead to catostrophic results, which lower plant quality. Common problems include phytotoxicity, delayed flowering, and stunted growth. The problems due to the overdose of some plant growth regulators are given below.

— *A-Rest (ancymidol)*. It is a safe chemical with limited reports of phytotoxicity. Phytotoxicity symptoms appear as necrotic spots on leaves of plugs and poinsettias. Problems are more pronounced with high temperatures (>70°F).

— *B-Nine (daminozide).* Phytotoxicity is not a problem with B-Nine. Because of its lower degree of activity, multiple applications may be required. These multiple applications can cause flower delay or reduced flower stalk elongation if applied too late.

— *Cycocel (chlormequat chloride).* Foliar sprays of Cycocel can cause a yellow spotting or discolouratoin of newly expanding leaves. Damage to the chloroplast causes the yellow spotting and symptoms are evident within 3 to 5 days after application. In general, problems occur when Cycocel rates exceed 1,500 ppm. Expansion of new leaves generally cover damaged leaves within a few weeks.

— *Bonzi (paclobutrazol) and Sumagic (uniconazole).* Both Bonzi and Sumagic are triazole type chemicals and have a high degree of activity. Over-application of these chemicals can result in leaf necrosis soon after application, darker green foliage with crinkled leaves, and severely stunted growth. Delay in flowering can also occur.

— *Florel (ethephon).* Overdoses of Florel are mainly associated with delayed flowering. Applying Florel to plants under stress (temperature or water) also can result in leaf yellowing.

Best management ractices for plant growth regulators do not end with the guidelines given above. For a total programme, growers should always monitor the effectiveness of applied treatments to assure that the treatments are working and to help "fine-tune" the amount of chemical needed. Monitoring treatment effectiveness may indicate that less chemical is needed. Always leave a few untreated controls mixed in with the treated population. This allows comparisons for effectiveness of the PGR treatment. For growth retardants, monitor plant growth for the treated plants regularly to know when plants begin to "grow out" of the treatment and to help in deciding if and when another application is needed. When used properly, chemical plant growth regulators are effective tools to help produce a high quality crop. When misused, they can reduce crop quality and increase production costs. Uniformity and consistency in application are crucial to attain predictable and desirable results.

8

Floriculture and Greenhouse Production

Ornamental plant production, is one of the fastest growing agricultural industries. It is a highly specialized and intensive form of agriculture. Unlike field or row crop agriculture, which bases production costs on acres, greenhouse costs are calculated on a per square foot of growing area or per plant basis. In addition, expenditures for greenhouse structures and plant material are considerable. These circumstances allow little room for error, so a great deal of planning is required before entering the greenhouse business. A person considering entering the greenhouse business must have a basic knowledge of propagation, plant nutrition, soil management, greenhouse structures, pest management, environmental control systems, marketing, employee relations and business management. Greenhouse operations often start as a single unit operated by a family. With modest-sized operations, experienced people function as "jack of all trades" while larger operations usually require a full-time and part-time staff.

Greenhouse production and management involves the propagation and production of herbaceous ornamental plants in environmentally controlled greenhouses, for later use in gardens, homes, and exterior or interior plantscaping. Floriculture greenhouse businesses produce and market their products to garden centers, floral shops, the mass market, and directly to the public.Proper plant nutrition is essential for successful greenhouse production of floricultural crops. As growers move towards substrates that do not contain mineral soil, micronutrient status of substrates and plants becomes more important.

Micronutrient Problems

Imapct of Excesses Micronutrient Application

Excessive application of micronutrients probably accounts for more micronutrient disorders in the greenhouse than does insufficient application. Excessive application of micronutrients, in addition to toxicities, can lead to micronutrient deficiencies. Deficiencies in this case are due to antagonisms between micronutrients during plant uptake. When two nutrients are antagonistic, a super-optimal concentration of one in the substrate (soil) will suppress plant uptake of the other.

A high level of iron in the substrate commonly causes manganese deficiency and to a lesser extent can suppress zinc uptake. Conversely, a high level of manganese in the substrate causes iron deficiency and also to a lesser extent, zinc deficiency. Super-optimal levels of copper cause zinc deficiency and conversely, high levels of zinc cause copper deficiency. Thus, it is possible to encounter deficiencies of iron, manganese, copper or zinc as a result of excess application of other micronutrients. These deficiencies can occur even when a normally sufficient concentration of the deficient micronutrient exists in the substrate.

Table 1. Common micronutrient antagonisms

High soil level of:	*Results in low plant level of:*
iron	manganese, zinc
manganese	iron, zinc
copper	zinc
zinc	copper

Many combinations of micronutrients are used in greenhouses. Each can be safe and effective when used in the role for which it was formulated. Excesses usually occur when multiple combinations of micronutrients are applied. This occasionally happens because some of the micronutrient sources are not obvious to the grower. Following are five sources or factors which provide or make micronutrients available:

1. Most substrates, whether commercially or self-prepared, contain micronutrients.
2. Most commercially formulated greenhouse fertilisers contain micronutrients. Fertilisers prepared by the greenhouse firm as an alternative to commercial products are often formulated with micronutrients. Generally, plants respond well to the combination of micronutrients in substrates and fertilisers.
3. Specific fertilisers are commercially available for use in "soilless substrates" that have higher micronutrient concentrations than the standard greenhouse fertilisers. The differences in micronutrient content between the standard and the soilless substrate formulations of one given fertiliser analysis are given in Table 2. The increase of micronutrients in the soilless substrate formulation ranges from 100 percent for iron to over 1000 percent for molybdenum. This third

source of micronutrients can be justified in a soilless substrate where the pH level is 6.0 or higher because the availability of micronutrients is strongly reduced in this situation.

Table 2.

The content of individual micronutrients in a general and a soilless substrate commercial formulation of 20-10-20 and the micronutrient concentration increase in the soilless substrate formulation.

	Content (%)		
Nutrient	*Standard*	*Soilless*	*Increase (%)*
iron	0.05	0.1	100
manganese	0.025	0.056	124
zinc	0.0025	0.0162	548
copper	0.0036	0.01	178
boron	0.0068	0.02	194
molybdenum	0.0009	0.01	1011

4. Research has indicated that, as is the case for organic field soils, the optimum pH level for organic (soilless) greenhouse substrates can be one pH unit lower than that desired for mineral-soil based substrates. The optimum pH range for soil-based substrates is 6.2 to 6.8 while for soilless substrates it is 5.6 to 6.2. When the pH of the substrate solution decreases, the availability of all micronutrients except molybdenum increases. Molybdenum availability decreases; however, deficiency of this nutrient is not known to be a problem in any floral crop except poinsettias. Thus, growing at a lower pH is equivalent to making an addition of micronutrients to the plant. Growers who maintain a pH level below 6.0 should consider using the standard fertiliser formulations rather than soilless substrate formulations that contain

higher levels of micronutrients, unneeded in this situation.

5. Often the four causes of increased micronutrient availability just discussed lead to excessive availability. Excess of one or more members of the micronutrient group can block uptake of another, bringing about a deficiency of the latter. It is then easy to mistakenly diagnose the total problem as a micronutrient deficiency. Without further information, correction is usually sought by applying a complete mixture of micronutrients. This makes the situation worse because the causal nutrients which are in excess become even higher in concentration. Even though the deficient nutrient is increased in the substrate, its uptake is not effectively increased.

Toxicity Correction

It is seldom possible to totally correct micronutrient toxicities. Vigilance should be exercised to prevent them. When toxicity does occur there are three steps which can be taken to reduce the problem:

1. The first and most obvious step is to stop application of micronutrients. Some fertiliser companies offer fertilisers without micronutrients. Otherwise, growers can formulate their own fertiliser without micronutrients.

2. The second step is to raise the pH of the substrate. The availability of all micronutrients except molybdenum decreases as the pH rises. Iron availability decreases tenfold when the pH level is raised one unit. Extreme shifts should be avoided. It is sufficient to move the pH level to the upper

end of the acceptable pH range for the crop. Three methods can be used for raising substrate pH:

a. A shift in the fertilisation programme from ammoniacal nitrogen (urea, ammonium nitrate, ammonium sulfate) to nitrate nitrogen (potassium nitrate, calcium nitrate) sources will bring about a gradual rise in pH.

b. Limestone may be applied to the substrate surface at a rate of approximately 1 lb/yd^3 per 0.1 unit rise in pH desired. This rate equals $^1/8$ tsp per standard 6 inch pot per 0.1 unit increase in pH and 1 $^3/8$ tsp per 6 inch pot for a 1.0 unit increase. These rates are for soilless substrates. Lower rates may suffice for soil-based substrates. Limestone reacts very slowly, thus two to six weeks may be required for a response. There are flowable suspensions of limestone that are more effective (faster reacting) than surface applications of ground limestone. Consult the label of commercial products for precise rates based on existing and desired pH.

c. For a rapid rise in pH, hydrated lime has been used. Caution should be taken to avoid contact with green tissues and neither limestone nor hydrated lime should be applied directly to substrates containing ammonium-containing fertilisers such as MagAmp or Osmocote. The high pH caused by these liming materials at the surfaces of such fertilisers can convert ammoniacal nitrogen to ammonia gas which is highly injurious to roots and foliage. To raise the substrate pH approximately 1 unit, one pound of hydrated lime should be mixed for five minutes in 5 gallons of water. Allow the mixture to settle overnight; and then apply only the liquid at the rate of one quart/ft^2 of bench area. This equates into 8 fl oz. per 6 inch pot. The substrate should be moist prior to applying the drench.

3. The third measure which might be taken for alleviating a micronutrient toxicity involves manipulation of antagonistic pairs of nutrients. If the micronutrient present in excess is a member of an antagonistic pair, make an application of the other member of the pair. For example, an excess of manganese, in addition to causing manganese toxicity, can result in iron deficiency. Application of iron alone will suppress manganese uptake and will increase iron uptake, thus alleviating both problems.

Diagnosing Micronutrient Status

It is important to diagnose the status of all micronutrients before undertaking corrective measures. Micronutrient disorders can involve one or more nutrients as well as combinations of toxicities and deficiencies. The presence of one micronutrient deficiency does not indicate that all other micronutrients are low. Application of a complete package of micronutrients in a situation where a deficiency/toxicity situation exists will increase the problem!

There are three systems for diagnosing nutrient status. The best diagnostic tool for micronutrients is foliar analysis. Visual observation of symptoms works but requires that damage be present. Most damage cannot be corrected. Commercial soil tests do not generally identify levels of all micronutrients. On the other hand, accurate tests and standards have been established for foliar analysis of all micronutrients. While the minimum and maximum critical foliar levels for micronutrients can vary for a few crops, these values do tend to be fairly standard for most crops.

Micronutrient Deficiency

Prior to adding micronutrients to correct a deficiency, check the substrate pH and make sure it is in the recommended range. If not, take steps to correct the pH before adding extra micronutrients. If there is still a problem then consider additions of the deficient micronutrient. There are three alternative methods of application for micronutrients:

1. Dilute concentrations may be applied in combination with macronutrients during each fertiliser application throughout the crop cycle. Sources, rates, and the final elemental concentration of each micronutrient are given in Table 3. This table will be helpful for those growers who formulate their own fertiliser and want to apply one or more but not all of the micronutrients.

Table 3.

Sources, rates, and final concentrations of one or more micronutrients with every liquid fertilisation

Micronutrient Source	*Weight of source/100 gal* *oz*	*grams*	*Final conc. (ppm)*
iron sulfate — 20% iron OR	0.13	3.7	2.00 Fe
iron chelate (EDTA) — 12% iron	0.22	6.2	
manganese sulfate —28%manganese	0.012	0.34	0.25 Mn
zinc sulfate — 36% zinc	0.0018	0.051	0.05 Zn
copper sulfate — 25% copper	0.0027	0.077	0.05 Cu
borax — 11% boron OR	0.03	0.85	0.25 B
solubor — 20% boron	0.017	0.48	

Contd....

Contd....

sodium molybdate —38% molybdenum OR	0.00035	0.01	0.01 Mo
ammonium molybdate— 54%molybdenum	0.00025	0.007	

When all of the micronutrients are desired most commercially prepared fertilisers can be used since they contain all micronutrients. When fertilisers are self-formulated, commercial products containing all micronutrients can be added into the fertiliser. Some of these products include Peters STEM, Peters Compound 111, and Miller's Mitrel M.

Table 4

Sources, rates, and final concentrations of micronutrients for a single corrective application of one or more micronutrients applied to the soil

Micronutrient Source	*Weight of source/100 gal*		*Final conc. (ppm)*
	oz	*grams*	
iron sulfate — 20% iron OR	4	113.4	62.0 Fe
iron chelate (EDTA) — 12% iron	4	113.4	36.4 Fe
manganese sulfate — 28% manganese	0.5	14.2	10.0 Mn
zinc sulfate — 36% zinc	0.5	14.2	13.9 Zn
copper sulfate — 25% copper	0.5	14.2	9.3 Cu
borax — 11% boron OR	0.75	21.3	6.25 B
solubor — 20% boron	0.43	12.2	
For soil-based substrates (>20% soil in substrate)			
sodium molybdate — 38% molybdenum OR	0.027	0.77	0.77 Mo
ammonium molybdate — 54% molybdenum	0.019	0.54	
For soilless substrates			
sodium molybdate — 38% molybdenum OR	2.7	77	77 Mo
ammonium molybdate — 54% molybdenum	1.9	54	

2. The second method of application calls for higher concentrations to be applied one time as a normal watering.
3. The third method involves a single foliar application of micronutrients. Sources, rates, and concentrations for foliar sprays are given in Table 5. Foliar sprays are very useful where root injury due to such factors as disease or a poorly drained substrate would reduce root uptake of nutrients. However, the greatest risk of plant injury exists with foliar application. Spraying should be avoided during the midday heat. Early morning, after sunrise, is an effective time for application. Plant uptake is enhanced by the increased drying time which occurs during the moist conditions in the morning. Nutrient uptake through the leaves is also greater in the light period than at night, thus making morning applications more desirable than evening sprays. Incorporate a recommended spreader/sticker into micronutrient sprays for more effective coverage. Use rates similar to those employed when applying pesticides.

Table 5

Sources, rates, and final concentration of the micronutrient for single foliar sprays for correcting micronutrient deficiencies

Micronutrient Source	*Weight of source/100 gal*		*Final conc. (ppm)*
	oz	*grams*	
iron sulfate	4	113.4	62 Fe
manganese sulfate	2	56.7	40 Mn
zinc sulfate	2	56.7	56 Zn
tri basic copper sulfate(53% copper)	4	113.4	159 Cu
sodium molybdate OR	2	56.7	57 Mo
ammonium molybdate	2	56.7	81 Mo

Control for Irrigation Water Used in Nurseries

High pH water and high alkalinity in water can be limiting factors in container production of greenhouse and nursery crops. An understanding of both is needed to accurately treat water with a high pH.

Substrate Solution and Water pH Factors

pH: A pH reading is a measurement of the hydrogen ion concentration of a solution (how acidic or basic a solution is), and readings range from 0 (most acidic) to 14 (most basic). Availability of micronutrients such as iron, manganese, zinc, copper, and boron and future plant growth can be reduced severely by high substrate and irrigation water pH (Figure 1). High pH water can cause salts to precipitate out of fertiliser stock tanks. High pH water can also reduce the efficacy of pesticides, such as Florel®. Therefore, growers should test and know the properties of their irrigation water.

Although pH 7 is considered "neutral" (not acidic or alkaline), 7 is not the optimum pH for irrigation waters or substrate solutions for nutrient availability and growth in container production. This is due to the substrate components typically used in nursery and greenhouse production.

The recommended range of irrigation water pH and substrate solution pH for production depends on the crop being grown. In general, pH should range from 5.2 to 6.8 for irrigation water and 5.4 to 6.3 for substrate solution. If the pH and alkalinity are high, your water may need acid treatment prior to use on crops.

Alkalinity: Alkalinity is a measure of a water's capacity to neutralise acids. Bicarbonate ions (HCO_3^-) from dissolved salts such as calcium bicarbonate

($Ca(HCO_3)_2$), sodium bicarbonate ($NaHCO_3$), and magnesium bicarbonate ($Mg(HCO_3)_2$); and carbonate ions (CO_3^{--}) from dissolved salts such as calcium carbonate ($CaCO_3$) are the major chemicals contributing to alkalinity in irrigation water. Hydroxide ions (OH^-) are but a minor contributor in most cases. Ammonia, borates, organic bases, phosphates, and silicates can also be minor contributors to alkalinity.

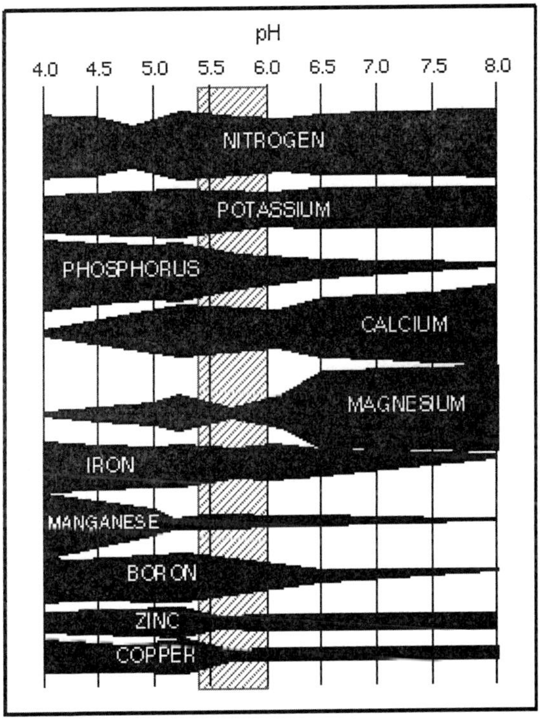

Figure 1. Influence of pH on the availability of essential nutrients in a soilless substrate containing sphagnum peat moss, composted pine bark, vermiculite, and sand. The pH range recommended for most greenhouse crops is indicated by slashed lines.

Total carbonates: Since bicarbonates and carbonates are the major components of water alkalinity, most laboratories assume that Total Carbonates (TC = carbonates +

bicarbonates) equals alkalinity. In most cases, this is a safe assumption. For most waters in the Southeast, bicarbonates account for more than 90% of all alkalinity present.

The term "alkalinity" should not be confused with the term "alkaline," which describes situations where pH levels exceed 7.0. Laboratory test results sometimes express alkalinity as a calcium carbonate equivalent, using milligrams per liter (or parts per million) of calcium carbonate (mg/L or ppm $CaCO_3$). The term "total carbonates" (TC) may also be used by some testing laboratories to refer to alkalinity of a solution. Some laboratories assume that all alkalinity is derived solely from bicarbonates (HCO_3^-) and will report alkalinity as bicarbonates using ppm (mg/L) or meq/L. To convert between these two units, use the following values: 1 meq/L HCO_3^- = 61 mg/L HCO_3^-.

Alkalinity establishes the buffering capacity of water and affects how much acid is required to change the pH. The following example may help explain the importance of alkalinity when trying to acidify water (Figure 2): Grower A has a water with a pH of 9.3 and an alkalinity of 87 mg/L HCO_3^- (TC = 1.42 meq/L). To reduce the pH of this water to 5.8, it takes 15.8 fl oz. of 35% sulfuric acid per 1,000 gallons of water. In contrast, Grower B has a water with a pH of 8.3 and an alkalinity of 378 mg/L HCO_3^- (TC = 6.20 meq/L). To reduce this water to a pH of 5.8, it takes 68.6 fl oz. of 35% sulfuric acid per 1,000 gallons of water. Even though Grower B's water is one pH unit lower than Grower A's, it takes more than four times more acid to lower the pH to 5.8 due to the differences in alkalinity. Both alkalinity and pH are must be considered when adjusting the pH of water.

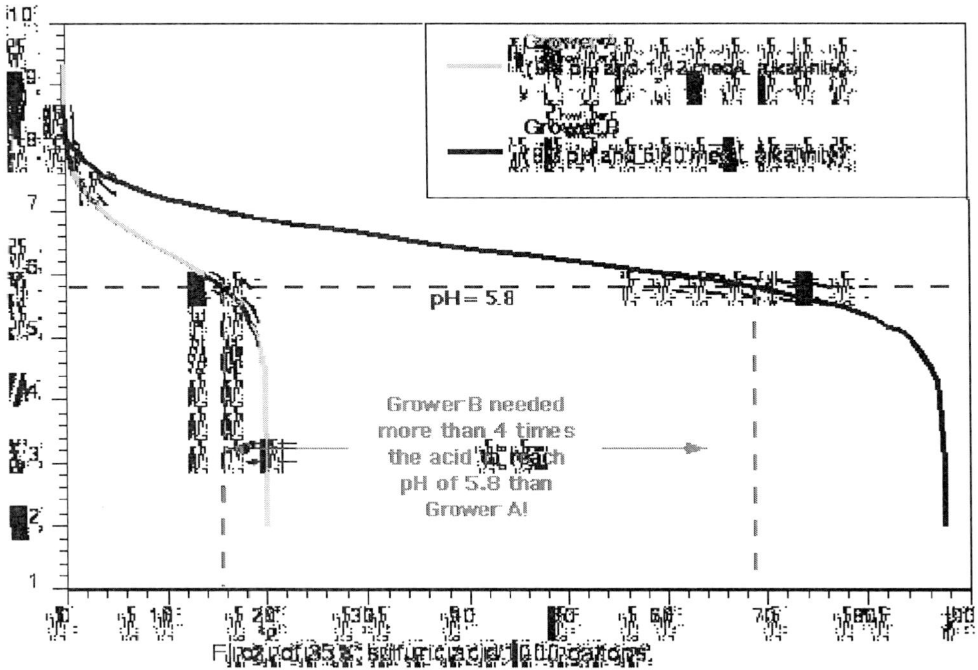

Figure 2. Titrations of two different waters with sulfuric acid. Notice that although the beginning pH of Grower A water is a full unit higher than Grower B water, it takes more than 4 times the acid (68.6 fl oz compared to 15.8 fl oz for Grower A water) to drop Grower B water to pH 5.8, due to the greater alkalinity in Grower B water.

Alkalinity can be a major problem in the North Carolina, especially in the coastal plains region. Levels of 2 meq/L and lower are safe for most crops. However, plug seedlings are more sensitive to alkalinity because the small volume of substrate provides little buffering against a rise in pH. Problems can occur in plug production if your water has more than 1.5 meq/L. If the alkalinity of your irrigation water is above 2.0 meq/L (or above 1.5 meq/L for a plug producer), you should consider injecting an acid to neutralise the bicarbonates (alkalinity), thus preventing an undesirable rise of substrate pH over time.

In-House Analysis of Water pH and Alkalinity

pH: Many relatively inexpensive ($30 to $250) pH meters are available that are accurate enough for greenhouse and nursery use. When selecting a pH meter, look for an accuracy of ± 0.1 pH unit and a range of 1 to 14. You should also be able to calibrate the meter (meter should have some type of calibration adjustment such as a set screw). Make sure to order standard pH solutions (usually pH 4.0 and pH 7.0) in order to calibrate your meter prior to use.

Alkalinity: Alkalinity is measured by titrating a water sample with an acid to an endpoint pH of about 4.6. A pH indicator dye (usually bromcresol green plus methyl red) is added to a known volume of water, and acid is added until the solution changes colour. With the bromcresol green plus methyl red dye system, the colour will change from green to pink. Most water sources acceptable for container production will have alkalinity in the range of 0 to 8 meq/L (0 to 400 ppm alkalinity expressed as $CaCO_3$). When looking for a test kit, this is the range that is needed. The level of accuracy does vary from kit to kit; ± 0.4 meq/L (20 ppm alkalinity expressed as $CaCO_3$) is accurate enough for most situations, but more precise kits are available. We have used Hach alkalinity kits #24443-01 (about $30 for 100 tests) and #20637-00 with good results. Although the second model is more expensive, it does have twice the accuracy (± 0.2 meq/L) and also comes with a digital titrator that can be used to measure other solution parameters (using different titrants and indicators) such as water hardness, chlorine, iron, nitrite, and sulfite concentrations.

Acidification Procedures to Neutralise Alkalinity

Acidification reduces the amount of bicarbonates (and carbonates) in water. Injecting acid into irrigation water neutralises alkalinity resulting in the formation of carbon dioxide and water:

$$H^+ \text{ (from acid)} + HCO_3^- \text{ (in the water)} \rightarrow CO_2 + H_2O$$

Sulfuric (H_2SO_4), phosphoric (H_3PO_4), nitric (HNO_3), or citric ($H_3C_6H_5O_7$) acid are commonly injected into irrigation water to neutralise water alkalinity. When deciding on which acid is best for your situation, evaluate

1. safety and ease of use;
2. the relative cost of the acid;
3. plant nutrient being injected with the acid (how much N, P, or S will be injected into your irrigation water); and
4. availability of the acid.

Safety: Citric acid, 75% phosphoric acid and 35% sulfuric acid are relatively safe to work with as compared to the 67% nitric acid. Nitric acid is very caustic and can cause serious injury to exposed tissue, especially eyes. Since nitric acid can also fume during handling, take care to avoid breathing fumes.

Avoid skin and eye exposure when handling any acid. Acid-resistant eye wear, gloves, and apron should be worn. Acids are corrosive and can damage clothing that is not immediately rinsed.

When mixing acid stock solutions, always add acid to a larger volume of water to create the stock. Never add water to concentrated acid.

Cost: In general, sulfuric and nitric are less expensive than phosphoric and citric. Citric acid is somewhat cost prohibitive for uses other than neutralisation of water used in pesticide sprays and in fertiliser stock solutions. Note the prices used in these calculations. Acid prices do vary with supplier and quantity. You may need to recalculate acidification costs using the cost of your acid.

Nutrients from acids: With the exception of citric acid, acids used for water acidification also supply a plant nutrient in conjunction with supplying H^+. The nutrient supplied can be beneficial to plant growth (if not supplied in excess), but it can also react with fertiliser salts in concentrated stock solutions or with pesticides if mixed into spray solutions.

Growers who acidify their water should adjust their fertilisation programme for the nutrient supplied by the acid. For example, if using phosphoric acid, make sure to reduce your P accordingly to account for the P supplied from the acid. When attempting to acidify waters very high in alkalinity, phosphoric acid may not be feasible. For example, if your water supply contains 6.0 meq/L of alkalinity and you used phosphoric acid to neutralise the alkalinity, over 126 ppm P (280 ppm P_2O_5) are supplied at each irrigation. This is an extremely high level of P considering a maximum of 55 ppm P (125 ppm P_2O_5) is recommended. Use another acid if more than this amount of P is being injected with phosphoric acid. If using nitric acid, account for the additional N supplied from the acid. Using 67% nitric acid to acidify water containing 6.0 meq/L of alkalinity would supply 67 ppm N at each irrigation, a significant quantity of nitrogen. Sulfuric acid treatment for 6.0 meq/L of alkalinity would supply 75 ppm S, more than sufficient sulfur for plant production (20 to 30 ppm S is suggested for most floriculture crops).

Acid recommendations: Citric acid is ideal as an acidifier for nutrient stock solutions and pesticide solutions, as it is much less likely to react with fertiliser salts or pesticides than the other three acids. However, the cost of citric acid makes it less desirable as an acidifier for large volumes of water used for irrigation and fertilisation.

For acidification of irrigation water, we recommend sulfuric acid over nitric acid due to its relative safety. Sulfuric is recommended over phosphoric acid for very alkaline water due to the higher cost of phosphoric acid and the possibility of over application of P. The sulfuric acid recommended is a battery electrolyte product named Qual® and can be purchased from most auto supply stores for about $10 per 5 gallons.

To accurately predict the amount of acid required to acidify to a given pH, both the pH and alkalinity of the irrigation water must be known. Table 6 outlines initial amounts of acid to inject to lower the pH to approximately 5.8. However, this table is only taking alkalinity into account (and does not account for the starting pH of a water sample), so fine tuning of the amount of acid added will be necessary.

Table 6. Acids commonly used to acidify irrigation water and their properties.

Acid	*Formulation and density (d) or formula weight (FW)*	*Amount of acid to add for each meq/L of alkalinity to result in a water pH of approximately 5.8*	*Concentration of nutrient provided by one fl oz. of acid per 1000 gallons water*	*Cost per meq/L per 1000 gal*	*Relative safety*
Citric acid (2-Hydroxy-1,2,3-propanetri-carboxylic acid)	99.5% (w:w) granular FW = 192.1	9.1 oz/1000 gals	none	\$0.59	can cause minor skin and eye irritation
$H_3C_6H_5O_7$ d = 1.21	50% (w:w) liquid	14.5 fl. oz/ 1000 gals	none	\$0.96	can cause minor skin and eye irritation
Nitric acid HNO_3	67% (w:w) liquid d = 1.42	6.6 fl oz/ 1000 gals	1.64 ppm N	\$0.26	use extreme caution; very caustic and dangerous; avoid contact with fumes as well as acid
Phosphoric acid H_3PO_4 d = 1.58	75% (w:w) liquid	8.1 fl oz/ 1000 gals	2.88 ppm P	\$0.44	slightly caustic; can cause skin and eye irritation as well as damage clothing
Sulfuric acid H_2SO_4 d = 1.26	35% (w:w) liquid	11.0 fl oz/ 1000 gals	1.14 ppm S	\$0.16	slightly caustic; can cause skin and eye irritation as well as damage clothing

When acidifying water, an end point of 5.8 is adequate to control substrate solution pH rise due to alkalinity in the irrigation water. Researchers from NC State University and Purdue University developed an Excel® spreadsheet that allows users to input their water pH and alkalinity, then select sulfuric, phosphoric, or nitric acid to use as an acidifying agent to reach a target pH or alkalinity. The spreadsheet modules calculate the nutrient additions from the acid injection and will report your acidification costs, if you input the price per gallon for the acid you wish to use.

Acid injection equipment: The injector used to add the acid into your water should be approved for acid injection by the manufacturer. Consult the manufacturer to select an injector for acid injection. Do not mix acid stock solutions with fertiliser stock solutions, as fertiliser salts could precipitate out of solution with certain acid/ fertiliser combinations. Use separate injectors for each solution. Acidic water will corrode galvanised piping and fittings over time. Be sure to check the integrity of greenhouse plumbing periodically.

9

Commercial Pansy Production

The pansy is a cultivated garden flower. It is derived from the wildflower called the Heartsease or Johnny Jump Up (Viola tricolour), and is sometimes given the subspecies name Viola tricolour hortensis. However, many garden varieties are hybrids and are referred to as Viola × wittrockiana. The name "pansy" also appears as part of the common name of a number of wild flowers belonging, like the cultivated Pansy, to the violet genus Viola. One or two unrelated flowers such as the Pansy Monkeyflower also have "pansy" in their name.

Pansies have been bred in a rainbow of colours, ranging from gold and orange though to purple, violet, and a blue so deep as to be almost black. They are quite a hardy plant, growing well in sunny or partially sunny positions. Pansies are technically biennials that normally have two-year life cycles. The first year they only produce greenery; they bear flowers and seeds in their second year of growth, and afterwards die like annuals.

Most gardeners buy biennials as packs of young plants from the garden center and plant them directly into the garden soil. Gardeners interested in rarer cultivars can plant seeds indoors in early November for plants ready in the spring. Under good conditions, pansies and viola are perennial plants, although they are generally treated as annual or biennial plants because they get very leggy and overgrown after a few years. The mature plant grows to 9 inches (23 cm) high, and the flowers are two to three inches (about 6 cm) in diameter.

Pansies are hardy in zones 4-8. They can survive light freezes or a little snow, but not for very long. In warmer climates, zones 9-11, pansies can bloom over the winter, and are often planted in the fall. In these climates, pansies have been known to reseed themselves and come back the next year. Pansies are not very heat-tolerant - once the temperature gets over a certain point they will become leggy and stop blooming.

Pansies should be watered thoroughly about once a week, depending on climate and recent rainfall. For maximum bloom, they should be given flowering plant food about every other week, according to the plant food directions. Regular deadheading can extend the blooming period.

Planting Pansies

Pansies are best used in boxes, pots, low borders or in mass plantings. They are often planted in between tulips and daffodils. Pansies are among the most enjoyable of the spring flower annuals or biennials. They bloom early with a wide range of colours, markings and sizes.

Pansies develop best when night time temperatures are in the 40's and day temperatures are in the 60 degree range. That is why they grow, develop, and flower profusely in March, April and May. Once hot weather comes they cease to grow and flower and are usually replaced with some summer annual.

Pansies are best planted in late winter even though they are often planted in late fall. They prefer a moist, rich soil that is well drained. Most flower beds need additional humus or organic matter for pansies. One to two inches of manure thoroughly worked into the soil will give excellent results.

Pansy plants may be purchased during late fall or early spring in bundles or in individual pots or as several plants per tray or pack. In the case of bundles, the plants should be placed in a bucket or tray with the roots covered by water for two to three hours before planting. After soaking, separate the plants carefully; the roots will not be damaged any more than is necessary. Plant them 12" apart and at about the same depth as they were grown.

When planting pansies in peat or fiber pots, pinch one or two holes in the bottom, and set them with the lip of the pot just under the soil surface. Water thoroughly to firm or settle the soil around the roots or the peat pot. Check the planting each day for water needs for the first two weeks. Afterwards, water thoroughly once per week. Try to water during the early morning hours.

Pansies have become the most popular annual for mid-fall to late-spring colour in the Southeast. Intensive breeding programmes that have selected for unique flower colours, large flower size, greater flower number, and temperature tolerance have led to many new and exciting cultivars to select from for use in the landscape. This leaflet was written to give growers production advice for pansies.

The modern pansy, *Viola X wittrockiana,* is thought to have derived from *Viola tricolour,* a native of central Europe. Although pansies are a perennial in cooler climates, they are grown as a cool season annual in the Southeast and rarely survive our summer heat. In the past 50 years new pansy colours such as shades of pink, rose, and orange have become available. Modern pansy breeding is largely concentrated in Germany, the United States, and Japan. From the late 1970's up to today, pansy breeding has concentrated on aspects of quality such as vigor, heat tolerance, and free flowering.

Flower Colour

Pansies have single blooms, each with five petals that are rounded in shape. There is a wide colour range for pansy flowers. Colours include red, purple, blue, bronze, pink, black, yellow, white, lavender, orange, apricot, and mahogany.

Pansies can be divided into pure-colour flowers and multicoloured flowers. Pure-colour varieties have a single colour on the flower and are called 'clear.' Multicoloured flowers that have a very dark blue/black centers are called 'blotched' or 'faced.' Some blotched pansies may have a different colour blotch than the usual dark blue/black face. Other multicoloured pansies have white or light coloured edges or have petals of differing colours; most of these two or three colour pansies also have a dark face.

Flower Sizes

There are three main categories of pansies, based on flower size:

1. Large — 3 ½ to 4 ½ inch-diameter blooms
2. Medium — 2 ½ to 3 ½ inch-diameter blooms
3. Multiflora — 1 ½ to 2 ½ inch-diameter blooms

There are well over 300 pansy cultivars available today. Most cultivars come in series (Table 1). A series has similar plant qualities such as plant size and heat tolerance. The individual members of a series each have different flower colours and sometimes have colour patterns that differ from each other.

Table 1. Major pansy series and cultivars with flower type, flower size, and number in each series.

Series	Plant type	Flower type	Flower size	No. in series or cultivar colour
Accord	F_1	both	medium	9
Alpenglow	O.P.*	blotch	large	Red
Atlas	F_1	clear	medium	2
Azure Blue	F_1	blotch	medium	Blue
Berna	O.P.	clear	large	Purple
Bingo	F_1	both	large	7
Characters	F_1	both	medium	14
Clear Sky	F_1	clear	medium	8
Colour Festival	F_2	blotch	large	Mix
Coronation Gold	O.P.	clear	large	Golden Yellow
Crown	F_1	clear	medium	12
Crystal Bowl	F_1	clear	multiflora	11
Delft	F_1	blotch	medium	Blue & White
Delta	F_1	both	medium	16
Faces	F_1	both	medium	16
Fama	F_1	both	medium	12
Giant Forerunner	?	both	medium	16
Happy Face	F_1	blotch	medium	7
Imperial	F_1	both	medium	15
Joker	F_2	blotch	multiflora	5
Lyric	F_1	blotch	medium	6
Majestic Giant	F_1	blotch	large	8
Maxim	F_1	blotch	multiflora	14
Medallion	F_1	blotch	large	6
Melody	F_1	both	multiflora	16
Padparadja	F_2	clear	multiflora	Orange
Paper White	O.P.	clear	large	White
Premiere	F_2	both	medium	6
Presto	F_1	clear	medium	8
Rally	F_1	both	medium	11
Regal	F_1	blotch	medium	13

Rhinegold	O.P.	blotch	large	Yellow
Roc	F_1	both	medium	11
Skyline	F_1	blotch	medium	8
Springtime	F_1	both	medium	18
Super Majestic Giant	F_1	blotch	large	9
Swiss Giants	O.P.	both	large	Mix
Ullswater	O.P.	blotch	large	Blue
Ultima	F_1	both	medium	4
Universal Plus**	F_1	both	multiflora	21

*O.P. = open pollinated variety.

**Replaces the Universal series.

Propagation

Pansy producers must decide whether to purchase pansy plugs or pansy seed. Seed may be grown as pansy plugs or in open seed trays, depending on time of marketing, labor availability, the number of pansies produced, and equipment and facilities available. For small operations and / or growers just starting into pansy production, purchasing plugs and concentrating on production of finished flats is a wise step. Only after gaining expertise in finishing seedlings should new growers consider purchasing seed rather than purchasing plugs.

Pansy propagation for the fall market must occur during July and August, months not conducive to optimum pansy seed germination and seedling development in North Carolina. A germination chamber and an efficient cooling system are needed for fall pansy production. There are many plans available for germination chambers. Many small and medium-sized greenhouse operations find it more feasible to purchase plugs for the early fall market rather than investing in a germination chamber or to tackle pansy germination in the summer heat. For late-autumn through early spring markets, pansy propagation is much easier due to cooler temperatures.

Germination Requirements

Seed selection is the first decision in pansy propagation. There are two basic types of pansy seed available, traditional seed that is not treated to enhance germination, and primed seed that has been physiologically treated to start the germination process. The advantages of primed seed include higher percentage germination, faster germination, and less sensitivity to high temperatures during germination. Primed seed cost more, but given the high temperatures in the Southeast during pansy propagation for autumn sales, primed seed should be used by growers lacking a temperature-controlled germination chamber.

Substrate physical properties for pansy germination include adequate aeration and good moisture retention. Germination substrates should be composed of finer (smaller particle size) components than substrate used for finishing bedding plants. A finer mix affords more moisture retention, but increases the possibility of overwatering and poor aeration.

Substrate chemical properties are as important as physical properties. A substrate pH of 5.4 to 5.8 is best for pansy production. Avoid substrates containing a large nutrient starter charge as pansy seedlings are sensitive to high concentrations of soluble salts. A substrate containing dolomitic lime for pH control (as well as to supply Ca and Mg), micronutrients, and a small amount of superphosphate is ideal. Phosphorus (P) levels in the germination substrate should be very low; high P levels cause seedling stretch.

Fertilisation using a liquid feed programme should begin after stem and cotyledon emergence. Use a fertiliser relatively low in phosphorus and low in ammoniacal nitrogen to prevent excessive stretch. Formulations with nitrogen derived from potassium nitrate [KNO_3], calcium nitrate [$Ca(NO_3)_2$], and magnesium nitrate [$Mg(NO_3)_2$] such as the basic-residue fertilisers listed in Table 2 are recommended. These formulations have the added benefit of supplying calcium and magnesium. Using these base residue fertiliser formulations allows pansy propagators to minimise the liming charge in the substrate and to use the fertilisation programme to maintain the appropriate pH.

Table 2. Quantities (ounces) of fertiliser or fertiliser salts to dissolve in 100 gallons of water to make solutions containing 50 to 250 ppm each of nitrogen (N) and potassium (K_2O). Select concentrations based on production phase and recommendations given in the text.

Fertiliser or salts	*Concentration of N and K_2O (ppm)* 50	75	100	250
Acid-residue sources			*oz/100 gallons*	
20-10-20*	3.34	5.0	6.7	16.7
20-9-20*	3.34	5.0	6.7	16.7
ammonium nitrate	1.23	1.85	2.5	6.15
+ potassium nitrate	1.50	2.25	3.0	7.5
+ monoammonium phosphate (20-10-20)*	0.54	0.81	1.1	2.7
Basic-residue sources			*oz/100 gallons*	
13-2-13 (-6Ca-3Mg)*	5.13	7.7	10.3	25.65
14-0-14 (-6Ca-3Mg)	4.76	7.14	9.5	23.8
15-0-15 (-9.5Ca-1Mg)	4.45	6.68	8.9	22.25
15-5-15 (-5Ca-2Mg)*	4.45	6.68	8.9	22.25
15-3-20 (-3.8Ca-1.9Mg)*	4.45	6.68	8.9	22.25
17-0-17 (-4Ca-2Mg)	3.92	5.88	7.83	19.6

Contd....

Contd....

potassium nitrate	1.51	2.28	3.03	7.58
+ calcium nitrate	1.76	2.64	3.52	8.8
+ magnesium nitrate	1.8	2.7	3.6	9.0
(13-0-13-6.6Ca-3.3Mg)				

*These formulations also contain phosphorus (P_2O_5).

Apply 50 ppm nitrogen feed every three to five days until the development of the first true leaves. At that point, increase the concentration to 100 ppm nitrogen, still applying every three to five days. Water with clear water in between feedings as needed.

Monitor the substrate pH during seedling production to assure it stays in the 5.4 to 5.8 range. A substrate pH above 5.8 can result in boron and iron deficiency; and high pH may lead to an increased incidence of black root rot, caused by the fungus Thielaviopsis basicola. If the substrate solution pH rises above 5.8, drench seedlings at 10 day intervals with 1 to 3 lb per 100 gallons of either iron sulfate [$FeSO_4 \cdot 7H_2O$] or aluminum sulfate [$Al_2(SO_4)_3 \cdot 18H_2O$] to help lower the pH. Lightly mist seedlings after application to prevent foliage injury from the drenches. If your irrigation water has a pH above 5.8, acidify it (at each watering and feeding) with 35% sulfuric acid to a pH of 5.4 to help lower the substrate solution pH. Continue these corrective treatments until the substrate pH drops and stays in the 5.4 to 5.8 range.

Temperature is an important factor for pansy germination and seedling growth. Optimum germination (fastest and greatest percentage) occurs at a constant 68 °F for both primed and non-primed pansy seed. After radicle (root) emergence (about 4 days after seeding) and during stem and cotyledon emergence, maintenance of cool temperatures (between 60 and 70 °F) is best for pansy seedling growth. Pansies do respond to DIF (difference between day and night temperature), so try to maintain a constant day / night temperature such as 68 °F rather than a positive (day warmer than night) DIF to reduce stretch. After the first true leaf appears (about 14 days after seeding) up to transplant into the finishing flat (approximately 6 weeks after seeding), temperatures should be reduced (if possible) to a constant 58 to 62 °F.

Light is not required for germination (during the first two to four days), but is needed after radicle (root) emergence to prevent excessive stretching. Between 1,500 and 3,500 footcandles of light are recommended after radicle emergence. Upon radicle emergence, many growers move seedlings from the germination chamber to the greenhouse, if they do not have adequate lighting in the germination chamber.

Moisture must be controlled during germination. Keep the relative humidity around the seed and seedling very high, but do not soak the substrate. Placing a thin layer of vermiculite over the seed will increase humidity around the seed during germination. Though relative humidity should be high during germination, excessive moisture in the substrate reduces oxygen, reduces germination, and reduces root growth. Frequent, light mist or fog applications are better than less frequent but heavier waterings that saturate the substrate. Reduce misting frequency and relative humidity as seedlings develop to 'harden' growth.

Seedling growth regulation is possible through DIF, as mentioned early. If a zero DIF (equal day and night temperature) does not control stretching adequately, apply a 5 to 10 °F negative DIF (night temperature of 64 to 68 °F and day temperature of 58 to 62 °F) for height control. When this level of temperature control is impossible, (often the case in the Southeast), consider using chemical growth retardants that are labeled for pansies to control seedling stretch (table 3).

Table 3. Chemical growth retardants labeled for use on pansies in the greenhouse.

Product	Stage of production	Application rate and method	Comments
A-Rest (ancymidol)	plugs or seedlings	3 to 9 ppm spray	(1.5 to 4.4 fl oz per gallon) apply beginning after the first true leaf is present
	plants in flats	8 to 19 ppm spray (3.9 to 5.8 fl oz)	make the first application when plants are 2" in diameter or height
B-Nine (daminozide)	plugs or seedlings	2,500 ppm spray (0.39 oz per gallon)	apply beginning after the first true leaf is present; weekly applications may be needed
B-Nine (daminozide) + Cycocel (chlormequat-chloride)	plugs or seedlings	1,000 ppm B-Nine (0.16 oz per gallon) + 1,000 ppm Cycocel (1.08 fl oz per gallon) applied as a tank mix spray	apply beginning after the first true leaf is present
Bonzi (paclobutrazol)	plugs or seedlings	1 to 5 ppm spray (0.032 to 0.16 fl oz per gallon)	use 3 ppm when two true leaves are present or use multiple sprays at 1 ppm
	plants in flats	5 to 15 ppm spray (0.16 to 0.48 fl oz per gallon)	use the higher rate in warmer weather; make the first application when plants are 2" in diameter or height; avoid late applications due to flowering delay and slow growth in the landscape after transplant
Sumagic (uniconazole)	plugs or seedlings	1 to 3 ppm spray (0.26 to 0.77 fl oz per gallon) make a 1 ppm spray when	use 3 ppm when three true leaves are present or two true leaves are present
	plants in flats	3 to 6 ppm spray (0.77 to 1.54 fl oz per gallon)	see comments for Bonzi

Many plug growers report that multiple applications at low concentrations offer more control in programming plug size than a single application at a high concentration.

Scheduling during seedling production depends on the environment and the production system used. Pansy plug production is a 5 to 7 week cropping period whereas seedlings produced in open flats are usually ready for transplanting 4 to 5 weeks after sowing.

Finishing Pansies

Timing for finished flats and pots of pansies should be based on a target market date and anticipated environmental conditions. Finished flats can be ready for sale in 5 to 9 weeks from transplant (total cropping time of 11 to 14 weeks). Count back from your projected market date to determine when you should have material ready for transplanting.

For autumn markets, many growers have plants available from mid-September through early December. In the piedmont and coastal regions of North Carolina, pansies are difficult to establish in the landscape prior to mid-October. However, given that customers are willing to purchase plants in September, growers usually target plant sales rather than survival.

Substrate characteristics that are recommended for the finishing containers are similar to those described for propagation. The major difference between the two is the degree of porosity; 'looser' mixes with a higher degree of aeration and more rapid drainage than germination mixes are recommended for finishing pansies. The substrate should contain liming sufficient to result in a pH of 5.4 to 5.8. Micronutrients and a small starter charge of phosphorus are also recommended.

Fertilisation during the finishing stage of pansy production is similar to that of the germination phase: do not overfeed pansies. A constant feed programme of 125 ppm nitrogen is sufficient for pansies, if leaching is minimal. If leaching is excessive (such as outdoor production subject to heavy rains), increase the constant feed to 175 to 200 ppm nitrogen. A weekly feed programme of 225 to 275 ppm nitrogen can also be used instead of a continuous feed programme. Rotate between a basic-residue fertiliser such as 15-3-20-3.8Ca-1.9Mg and an acid residue fertiliser such as 20-10-20 to assure adequate supplies of macronutrients and to help maintain a constant substrate pH.

Temperatures for finishing pansies are similar to those used in seedling production. Best growth and development is best achieved with a daily average temperature of 58 to 62 °F. As mentioned earlier, pansies do respond to DIF, and a zero or negative DIF is preferred for height control over a positive DIF. Pansies can be grown at even cooler temperatures (night temperature of 48 °F) without damaging the plants, but development will be slower.

Light levels should be as high as possible, but temperature control usually requires shading of about 30 to 50% during late summer and autumn to reduce the heat load on the plants. Remove shading and increase irradiance as soon as ambient temperatures allow to promote better plant growth and flower development.

Water plants prior to wilting, but allow the substrate to dry out between waterings. Excessive moisture results in poor root growth and could increase the incidence of root rots. However, allowing the substrate to dry to the point of plant wilting could concentrate salts around the roots, resulting in root damage. When watering, try to minimise leaching (maximum of 15% leaching fraction) to reduce nutrient runoff from your crop.

Plant growth regulation is best achieved through a zero or negative DIF. A day / night temperature regime of 55 °F/65 °F is very effective in reducing excessive stretch, but is a rarity during autumn. Do not apply growth retardants to transplants until they exhibit active growth. Avoid late applications of growth retardants (especially of Bonzi and Sumagic), as they may delay flowering and slow growth once plants are transplanted into the landscape.

Production Problems

Insect and Pests

There are a variety of pansy pests, some of them with the potential to be serious problems (Table 4).

Table 4. Insect and related pest problems of pansies during production, their botanical name, and NC State University Department of Entomology Ornamentals and Turf Insect Notes that address each pest.

Pest	*Botanical name*	*Insect note(s)*
Black cutworm	Argotis ipsilon	#7
Foxglove aphid	Aulacorthum solani	#38, #103, #104
Fungus gnats	Lycoriella spp. & Bradysia spp.	#29, #103
Green peach aphid	Myzus persicae	#38, #103, #104
Greenhouse whitefly	Trialeurodes vaporariorum	#10, #103, #104
Pansyworm	Euptoieta claudia	#7
Shore flies	Scatella spp.	#103
Silverleaf whitefly	Bemisia argentifolii	#103, #104
Slugs	Deroceras spp. & Lehmannia spp.	#22, #91
Variegated cutworm	Peridroma saucia	#7
Western flower thrips	Frankliniella occidentalis	#72, #103, #104
Yellow woollybear	Spiosoma virginica	#7

Growers should contact their County Extension Centers to request the suggest Insect Notes dealing with each pest. These insect notes and the current edition of the "N. C. Agricultural Chemicals Manual" will offer the most up-to-date control measures for each pest.

The major pests listed in Table 4 that growers should be scouting for are printed in bold type. The other pests listed have the potential to become production problems, but are usually not as serious.

Diseases: Pansies are susceptible to a number of serious diseases in both production greenhouses and in landscape beds. Listed below are some of the more common pansy diseases in North Carolina along with prevention and control measures. Consult the current edition of the "N.C. Agricultural Chemicals Manual" for the most up-to-date listing of fungicides labeled for use on pansies.

— *Damping-off of seedlings:* During production of pansies, damping-off by Pythium spp. can be a major disease problem. The best control of any disease is to completely avoid its occurrence. Every bedding plant grower should thoroughly understand damping-off, factors that favor its development, and potential sources of contamination. Damping-off must be avoided through a complete, continuous sanitation and cultural programme. Fungicides should be used to control damping-off only as a last resort. Most fungicides used to control damping-off also cause some injury to seedlings. The younger the seedling, the more severe the injury. Fungicide rates higher than recommended must be avoided. If damping-off is a continual problem, much effort should be made to determine the source of the pathogen and correct the sanitation or cultural problems rather than continue to use or increase the use of fungicides. Fungicides effective in controlling damping-off due to Pythium spp. include etridiazole (Truban) and metalaxyl (Subdue).

— *Black root rot:* Black root rot, caused by the soilborne fungus, Thielaviopsis basicola, can also be very serious on pansy. This fungus attacks the fine feeder roots. Infected roots are black due to the presence of the fungus. Thielaviopsis basicola also causes a root rot on Helleri holly and vinca.

This fungus is common in soils across North Carolina and it is active over a very wide temperature range. The fungus infects the feeder roots and gradually kills the entire root system.

Black root rot has been a serious problem in pansy production in the Southeast for the past five to seven years. It appears to be related to production in August and September when it is too hot for pansy plants and they are "stressed."

Control measures for black root rot during production include:

1. Avoid excessive heat stress, especially on young seedlings.
2. Avoid other stresses, such as high substrate pH that can lead to micronutrient deficiencies (mainly boron and iron), excessive soil moisture, or excessive salts.
3. Preventative drenches with labeled fungicides such as 1-[1-[[4-chloro-2-(trifluoromethyl) phyenyl]-imino]-2-ropoxyethyl]-1H-Imidazole (Terraguard 50

WP) and thiophanate methyl products such as Domain or Cleary's 3336 may help.

— *Botrytis blight:* Botrytis blight is a common but only slightly damaging disease of pansies caused by Botrytis cinerea. Botrytis cinerea is an airborne fungus that attacks almost any flower and dead, dying or damaged plant tissue. During production, the greatest losses due to botrytis blight occur late in production. Occasionally, this fungus also can attack germinating seed or seedlings, particularly if they are injured or excessively crowded. High rates of fertilisation, death of lower leaves, low light intensity, frequent watering, early flower production, and crowded plants all favor botrytis blight development. The development of botrytis blight is also favored by warm temperatures and high humidity. Most growers try to manage botrytis blight with a combination of cultural practices, sanitation, and timely fungicide applications. Some effective management tools include:

1. Keeping the relative humidity below 85%, particularly at night.
2. Keeping the spore load down in the greenhouse through good sanitation.
3. Good air movement in the greenhouse to reduce humidity pockets.
4. Proper ventilation of the greenhouse.
5. Greenhouse covers that reduce spore production.
6. Sunny weather with lower relative humidity.

7. Crop mix made up with plants less susceptible to botrytis blight.
8. Ship plants at their peak of quality.
9. Timely applications of effective fungicides. Fungicides with potential resistance problems include: thiophanate methyl, iprodione, and vinclosolin. Fungicides with no resistance problems include: chlorothalonil and mancozeb.

— *Leafspot diseases:* Pansies are susceptible to several leafspot diseases, but the most common ones in North Carolina are anthracnose (caused by the fungi Colletrotichum gloesosporiodes and Colletrotichum violae-tricolouris) and scab or spot anthracnose (caused by Sphaceloma violae). The leafspots vary in colour from white to brown to black and often have a water-soaked margin. The spots may or may not have concentric rings and spore producing structures. While leafspot diseases are fairly common on pansy, they seldom cause much damage. These diseases are best controlled through proper sanitation, such as removal of plant debris.

— *Additional diseases of pansy:* Three rust diseases (caused by Puccinia ellisiana, Puccinia violae, and Uromyces androponis) plus a seed smut (Urocystis kmetiana) have been reported on pansy, but rarely are a major problem. Other diseases reported on pansy include root rot caused by Aphanomyces cladogamus; leafspot diseases caused by *Alternaria violae, Cercospora violae, Phyllosticta rafinesquii, Ramularia agrestis,* and *Ramularia lactea; downy mildew* caused by *Peronospora violae; damping-off* caused by *Rhizoctonia solani;* Southern blight caused by *Sclerotium rolfsii;* powdery mildew caused by

Sphacerotheca humuli; and root knot nematodes, *Meloidogyne spp.*

Prior to treating for these or any disease, it is essential to have the disease organism properly identified. Contact your County Extension Center for sample submission procedures.

Nutritional Disorders

Pansies are relatively free from nutritional disorders, when grown at the proper pH. However, when the substrate pH is allowed to climb above 5.8, micronutrient deficiencies can be a problem. Magnesium (Mg) deficiency can also be encountered, if the pH falls too low, or if calcium levels are too high with respect to Mg levels. Routine foliar analysis should be conducted to assure that nutrient content in the foliage is within recommended ranges (Table 5).

*Table 5. Recommended foliar nutrient concentrations for pansies.**

Nutrient	*Adequate range*
Nitrogen (N)	2.5 to 4.5%
Phosphorus (P)	0.25 to 1.0%
Potassium (K)	2.5 to 5.0%
Calcium (Ca)	0.6 to 3.0%
Magnesium (Mg)	0.25 to 0.75%
Sulfur (S)	0.2 to 0.7%
Sodium (Na)	0 to 0.5%
Iron (Fe)	30 to 300 ppm
Manganese (Mn)	25 to 300 ppm
Zinc (Zn)	20 to 100 ppm
Copper (Cu)	5 to 40 ppm
Boron (B)	20 to 80 ppm

*Ranges are averages from three commercial analysis laboratories.

Boron deficiency

The symptoms of this deficiency are very specific. The younger, developing leaves are small, thickened and puckered. In many instances, the main shoot will stop growth completely (abort), and lateral shoots will attempt to expand, developing the thick, puckered leaves mentioned above.

The first stage of treatment should involve reducing substrate pH to 5.4 to 5.8 to make boron more available to the plant.

The second step in treating boron deficiency would be to apply a substrate drench of borax at $^1/2$ oz per 100 gallons or use solubor at $^1/4$ oz per 100 gallons. Lightly mist off foliage after the application, as boron solutions can burn leaves.

Prior to the boron drench, check the levels of calcium and magnesium in the substrate. Calcium tends to tie up boron, especially when the calcium to magnesium ratio is too high (greater than about 5 : 2, Ca to Mg). If Ca is out of proportion to Mg, include 1 lb per 100 gallons of Epsom salts ($MgSO_4 \cdot 7H_2O$) in your boron drench.

Do not apply more than two boron drenches during production, as excessive boron will also cause problems on pansies. Unfortunately, plant recovery from boron deficiency is a slow process. It may take two to three weeks before normal growth is resumed after a boron application.

Iron deficiency

Symptoms of iron deficiency are interveinal chlorosis (yellowing) of primarily the youngest leaves, followed by marginal burning in severe cases. As with boron deficiency, the first step in treating iron deficiency is assuring the substrate pH is within the recommended range. If the pH is too high, follow the recommendations given earlier to lower it to 5.4 to 5.8 using iron sulfate. Not only will this treatment lower the pH, it will also increase the iron supply in the substrate solution. If further treatment is needed, use a foliar spray of 10% iron chelate at 4 oz per 100 gallons.

Magnesium deficiency

Symptoms of magnesium deficiency on pansy are interveinal chlorosis of the newly matured leaves followed by general yellowing of the leaves beginning at the margins. Marginal necrosis can follow in severe cases.If magnesium deficiency is suspected, check the Ca to Mg ratio. If it is greater than 5 : 2, apply a substrate drench of 2 lb Epsom salts per 100 gallons of water. Do not make applications more than once every four weeks. If multiple applications are needed, be sure to monitor both foliar and substrate levels of Ca to assure that the Mg applications do not cause Ca to become deficient.

Our understanding of the growing requirements for pansies has improved. We now know that the environmental requirements for this species are the most stringent of any commercial bedding plant crop. Pasies are best considered a specialty bedding crop. Success with this crop requires growers to reevaluate their cultural programme and make the necessary greenhouse adjustments. Those who implement the growing programme changes will see how trouble-free and profitable this crop can be. The secret to growing and maintaining high quality plants is maintaining an optimal root environment. To accomplish this, you must undertake simple but important changes in your growing programme.

Marketing

Because of the increasing popularity of pansies, many growers have become interested in producing their own crops for several markets. In the spring and fall months, homeowners who plant pansies around their residences are the largest market. One additional market in the fall is commercial landscapes. Landscape contracting businesses purchase a significant number of pansies in the fall for their residential and corporate clients. While the market for pansies has grown dramatically in the recent past, growers just entering the business should first develop a plan and determine how much of the market they would like to serve.

The two market windows in which pansies are sold are fall and spring. The fall market is the larger of the two as many gardeners and professionals prefer to get several months of blooms from their plants by establishing them in the fall rather than the spring. The early spring season for bedding plants, February through April, may offer some potential for pansy plant sales. The market also contains two distinct groups who will purchase the plants: consumers or end-purchasers and retailers. The consumer or end-user market is probably the largest market segment. Householders may purchase a significant number of pansy plants for their home landscapes. Small businesses, such as real estate firms or banks, may landscape their own properties and purchase pansies in the fall or spring.

The second portion of the market are customers who will resell the pansies. Some pansy growers produce plugs for other growers to finish. Consider, too, landscape contractors and the types of plant material they need in the fall. Many of them will purchase pansies for fall planting. Other market segments include garden centers, discount stores, florists, and other retailers.

10

Commercial Floriculture in Asian Countries

Floriculture has emerged as a profitable agri-business option the world over in recent years, particularly in the developing nations. World floriculture trade totals more than US$ 6,800 million. The Netherlands alone account for nearly half of this trade and countries like Colombia, Italy, Denmark, USA, Belgium, Israel, Costa Rica, Germany, Canada, France, Spain, Kenya and Ecuador all have a share of more than US$ 100 million. Asian countries are also entering the international trade and making their presence felt. Thailand, Malaysia, China, Singapore and India are leading the Asian contribution. However, while their present share of the trade appears insignificant, the growth of these countries in recent years has been very impressive. For cut flowers with a trade of US$ 3,700 million Thailand has a share of 2%. India which now has a share of only 0.3%, had a growth rate of 563% in the 1991-95 period. The other country with impressive growth rate has been Malaysia (259%) over the same period.

Among the growing markets (cut flowers, cut foliage, live plants and cuttings) Germany with a share of US$ 2,000 million was the world's biggest market, followed by

USA with a share of US$ 831 million. Japan, though seventh in ranking with a trade worth US$ 314 million, emerged as the fastest growing export market during the 1991-95 period indexing a growth of 162%. Consumption in these three countries are characterised by the domination of middle class population. The world production of floriculture products at a very conservative estimate is valued at $ 20 billion. Japan with a production value worth US$ 3,736 million leads, and with the Netherlands (US$ 3,558 million) and the USA (US$ 3,270 million) account for nearly half of the world total.

In terms of total area in production, Asia and Pacific with an area of 134,000 ha cover nearly 60% of the total world area. A significant feature of the production area statistics is that in most Asian countries, only a small share of the area contributes to production for world trade (exports). This is because either only a small percentage of the area is under protected cultivation (as in China and India with large total areas) or the production is mainly for the domestic market (e.g. Japan).

Global production of floriculture products will continue to expand, especially in new producing countries in Asia, Africa and Latin America, though the traditional large producing countries like the Netherlands will also work towards higher productivity per worker and per unit area. In general, however, the trend in production of cut flowers which is going to be very labour intensive, will tend to be located in areas with lower labour and other production costs. The production strategies will require reduced costs so that the profitability improves. For the purchaser it is not important where the flowers come from. They are mainly interested in price, quality, freshness and variety.

For the growing flower exporting countries of Asia, one of the major barrier in increasing profitability is import duty. Around the world, duties are used to protect the local flower industry from unfair competition. The only exception being Japan, among the leading floriculture markets, which charges no duty as yet. But the total imports of cut flowers into Japan are less than 7% of their trade. It remains to be seen, if Japan will continue to have no non-tariff barriers or restrictions on trade, when China, assisted by Dutch technology and quality propagating material from around the world, increases its exports to Japan, particularly considering that China itself has a duty of nearly 53%.

For Asian countries, another hampering factor is the soaring cost of air freight, which may take away the competitive advantage of low cost land, labour and year round growing climate. Various international carriers no longer prefer flowers because they have alternative and more lucrative 'exotic' cargo such as computer chips.

The cut flower trade worth $ 3.7 billion is likely to increase by around 10% annually. The share of the developing countries of the total trade has consistently been around 20%. During the last five years there has been a consistent growth in the number of producing countries particularly among the developing nations in Asia, Africa and Latin America. In general, the world supply is increasing more than demand, resulting in reduced prices per stem. With a highly perishable product like flowers, even a small over-supply can have a dramatic effect on prices. This will require a greater attention to be paid to identifying crops ideally suited for a particular location, as well as product diversification. One must critically analyse the strengths, weaknesses, opportunities and threats before embarking on expansion of the production/trade programmes. Greater attention to high value crops, though traded in lower volumes, will increasingly become more important.

Orchids, while not a major crop in most regions, have a special place in floriculture. Another crop demanding better attention in world trade has been Anthurium. Its world trade is second only to orchids among tropical flowers. Its popularity is increasing, particularly in Asia.

Most Asian countries have potential for growing high value crops and need to devote larger areas mainly to meet the huge growing demand in Asian markets, where their freight rates will also be lower. Countries in Asia also require identification and commercialisation of certain native tropical plants to cater to niche markets in Europe and the USA.

As importers are trying to import from the cheapest country, the Asian growers could easily lose their large European market, in view of the strong competition from growers from Africa and Latin America, who may have lower freight costs and no or little import duty to pay. In addition the growers in Holland, with the advantage of producing right in the heart of the most developed market in the world, with low or no transportation cost or duty, would always enjoy the best marketing opportunity. The Japanese and the US markets with relatively low import volume will continue to satisfy their supplies from nearby countries.

The other growing Asian markets like Singapore, Hongkong, South Korea, etc. are hardly in a position to accept large volumes. One also needs to consider the emerging new markets in Middle East countries and Russia. The best option however, would be for all the major flower growing nations in the region to concentrate on developing their own domestic market. With increasing consumer wealth, population concentration in major urban areas and improved living standards, India, China, South Korea among others appear to be potentially important flower consuming countries for the future. In these countries, as also in others, most flower consumption occurs around major holidays, like New Year, Valentine's Day, Memorial Day, Mothers' Day, Fathers' Day, Christmas, and in wedding seasons. The governments in these countries are making sincere efforts to develop the philosophy of people with regard to flowers through strong promotional efforts.

Foreign investors are being encouraged to help develop their floriculture industry through joint venture projects. All these efforts will hopefully lead to making available quality produce for the consumers. The governments will need to invest more on developing the required infrastructure, particularly for marketing like auction centres etc. and increase access to new and improved technologies and quality inputs. The efforts to develop the competitiveness in the export markets, through providing subsidies for capital investments, air freight etc. should also continue alongside.

Evolution of Cut Flower Market

Four decades ago, demand for cut flower by consumers around the world was satisfied by local cut flower production. In Europe, per capita consumption was significant, and consumer culture required a large supply of cut flowers for gifts, occasions, and everyday use. As a result, cut flower production in Europe was sizable. Gradually as developed throughout this region, it became possible to distribute cut flowers grown in southern areas of Europe to northern areas of Europe. Consequently, the European flower industry began to extend its boundaries for cut flower production and along with this expansion grew the influence of the European flower industry. This background history could be considered the beginning of commercial floriculture as we know it today.

When the world energy crisis occurred in 1973, the marketing plan for distributing cut flowers grown in different European countries to Holland for sale through the Dutch flower auction and back to markets throughout Europe became a significant production opportunity for southern European cut flower growers.

Increasingly larger quantities of cut flowers were grown in southern Europe to meet the demand for cut flower sales through Holland. Flower growers in the southern regions had a price advantage over growers located in northern regions because cut flower production was more expensive for northern growers during the winter season due to increased energy costs required to obtain quality flowers in controlled temperature greenhouses.

Then, competition for southern European cut flower growers intensified when Israeli cut flower growers, who were located further south entered the market with product to be sold through the Dutch flower auction. Israeli growers had the production advantage of being further south where they could produce cut flowers in open fields or plastic tunnels year round, eliminating most of the overhead expenses for greenhouses and heating systems. But in order to develop a potentially lucrative export cut flower industry for themselves, the Israelis needed to address limiting factors to their success. The two main limiting factors were transportation costs to Europe and a water shortage if production were to expand.

Solutions to these limiting factors were found for Israeli growers. In the case of transportation costs which offset growers cost advantage in terms of energy compared to growers in southern Europe, the government provided transportation subsidies which have reduced the costs to the growers to ship their cut flower product to Europe, thereby maintaining a competitive cost advantage over European growers. As for the water shortage, research on irrigation systems that would conserve water usage was applied to production systems for cut flowers.

Through the 1970's, the activities of the European flower industry had begun to influence cut flower production and sales beyond the borders of Europe. Cut flower sales through the Dutch flower auctions had gained a share of the United States market. This was achieved by promotion activities in the USA supported by the Holland Flower Council which encouraged Americans to purchase more cut flowers for gifts, occasions and everyday use, similar to consumer habits in Europe. Most of the flowers sold to the USA through the Dutch flower auctions are shipped to the USA by air through New York.

Simultaneously, Miami, USA, was being developed as a key import distribution base for cut flowers being grown in Columbia, South America and shipped north. This caused considerable competition for local cut flower growers in the USA. Manufacturers and suppliers from the European flower industry were quick to find opportunity in this situation.

Not only were South American cut flower growers purchasing varieties from Europe but flower growers from the USA were persuaded to invest in production systems and equipment from Europe in hopes of becoming more efficient producers like the Dutch growers who had once faced competition from southern European growers. As a result, the United States flower industry owes a significant share of its growth in terms of promotion and sales and improved production systems to the influence of the European flower industry.

It is worthwhile to mention that the Israeli flower industry has become a formidable competitor of the European flower industry. Israeli cut flower producers ship significant quantities of product into the USA market via both New York and Miami. This compensates Israeli producers for the reduction in cut flower sales to the European market which is increasingly being supplied by flower growers from regions in Africa, especially Kenya. Also, Israelis have been successful in selling their production equipment and varieties to flower growers in other countries.

Continuing to advance in the 1980's, the European flower industry began seeking further opportunity and expansion in Asia by 1985. Japan's bubble economy was starting to inflate and discretionary income spending by the Japanese was rising. European flower imports made headway into the lucrative market in Japan. Within a few years, as economies in Korea, Taiwan, and Hongkong strengthened, the European flower industry moved into these markets with their cut flower exports as well.

Since the early 1990's, the European flower industry, as a worldwide leader in commercial floriculture, has been impacting the rest of Asia with cut flower imports from Holland and sales of flower varieties, production equipment, and technology for new production operations in Asia. Israeli cut flower producers, manufacturers, and suppliers have followed but, one step behind.

The main difference between the European flower industry and the Israeli flower industry is that the European flower industry enters their new markets by launching aggressive marketing campaigns which call attention to the quality and image of Dutch flowers. These campaigns stimulate demand by new consumers for their cut flower products. So far, the Israelis have not particularly created an image for end consumers of Israeli flowers. This difference is one of the factors which contributes to why the European flower industry is the world wide leader in commercial floriculture.

Initially, commercial floriculture production in southeast Asia was developed because of increasing need for low cost flowers by the European cut flower market place. European flower traders identified commercial floriculture production in southeast Asian countries as a source of supply. Ironically, Dutch auctions often served to re-distribute this product to the Japanese market. By the mid to late 1980's, Dutch importers/exporters had begun selling floriculture product in Japan. With economies expanding, the "little tigers", i.e. Taiwan, Korea and Hongkong were the next Asian targets with market needs for floriculture products from Europe and potentially from other Asian countries which could produce floriculture products less expensively.

The development of the commercial cut flower industry in Asia has been unlike that of Israel, African countries, south and central American countries. In the latter regions, cut flowers have been a product produced mainly for export with no thought of a potential domestic market. On the other hand, in Asia, whereas cut flowers were initially produced for export, the market potential has rapidly changed to include opportunities for supplying the local market as well. This unique development is on account of the rapid strengthening of economies in the region, high population densities, and the consumer perception which has been promoted heavily by the European flower industry that the use of fresh flowers in one's every day life represents an improved, quality lifestyle.

Today and in the future, the potential for commercial floriculture expansion in Asia, including production for domestic and export sales of cut flowers, is greater than ever before. The elements for success needed to transform the Asian flower industry into the worldwide leader of commercial floriculture have been implied by reviewing the evolution of the world cut flower market place.

Opportunities for Cut Flower Development

There is considerable opportunity in both export and domestic markets for cut flowers from southern Asia. In case of export markets, it seems reasonable to set short, middle, and long term goals with respect to entering these markets. Simultaneously, efforts to develop domestic markets within and among countries in southeast Asia should be undertaken.

Establishing Export Markets

Already, cut flowers have been exported from southeast Asia to other countries in Asia, especially Japan, Korea, Taiwan, and Singapore. The advantages in marketing product to these countries is that they are relatively near, shipping requirements in terms of postharvest storage have been minimal, though this situation may not persist, and often contacts to find buyers have been made only through relative or previous business relationships.

In the case of Japan, it is well known that custom regulations and quality requirements are among the strictest in the world, though the price fetched for product is high. This incentive is attractive despite the difficulty to achieve success in marketing cut flowers to Japan.

In terms of distribution to these countries, it is important to examine what are the practical transportation routes to these countries from southeast Asia. It could be that developing a transportation hub in Bangkok or Singapore would be convenient for shipping cut flowers from southeast Asian countries, including cut flowers from Yunnan province in China, to countries in northeast Asia.

Numerous commercial flights already are in place regarding this possible shipping pattern for cut flowers from the region. The transportation concept of a shipping hub would imply that initially all cut flower production intended for export to northeast Asian destinations would first be shipped to the hub, e.g. Bangkok, and then redistributed to specific destinations in Japan, Korea, Taiwan, and Singapore. This routing for northeast Asia could be extended to Australia and New Zealand.

This centralised shipping hub would require expanding postharvest handling and storage facilities at the airport terminal and increasing freight forwarding activities for cut flowers in Bangkok. At the same time, shipping procedures and storage requirements for cut flowers to be transported from southeast Asian countries would be streamlined by the ability to focus on transportation requirements to Thailand rather than an array of destinations. In the short term, this could serve as a way to shorten the time to improve distribution of cut flower product from all southeast Asian countries to these lucrative northeast Asian markets.

At first consideration, it may seem unlikely that cut flowers exported from southeast Asia could penetrate the markets in North America or central Europe as these markets are already well supplied with cut flowers from other regions such as south and central America and southern Europe, Israel, and Africa. However, in the case of North America, the entry points for cut flower imports are mainly New York and Miami, leaving open the opportunity for establishing an import entry point in Vancouver or San Francisco, especially for cut flowers shipped from a Chinese port, i.e. Kunming. Recently, with the return of Hongkong to China, there has been an exodus of Chinese from Hongkong to Canada establishing residence in North America. Already the San Francisco area in the USA is home to many ethnic Chinese.

Due to the high density of Chinese in these two western entry points to North America, it is highly conceivable that Chinese flower importing companies could proliferate if the opportunity were available. This was certainly the case with Miami which already housed a large population from south and central America when flower imports from Columbia, Ecuador, and Costa Rica began to enter the USA market on the east coast.

Until now, domestic producers of cut flowers in California on the west coast of the USA have been holding up against imports from south and central America but, would surely retreat if low cost, high quality imports from southeast Asia via China were available. The reason for setting this export goal as "mid-term" is that airline shipping routes from China to these cities are not as well established at this time compared to routes from Bangkok to northeast Asian destinations.

It might be proposed to establish Kunming in China, which is currently being developed as the center for cut flower production in China, as the transportation hub for distributing flowers from southeast Asia to North America.

Another mid-term export goal would be to market flowers from southeast Asia to central Europe. This could be achieved, transportation wise also through Bangkok, which already has airline clearances for many central European cities, including Amsterdam.

Another factor in identifying mid-term export market goals is that both north American and central European markets are already well supplied with an abundance of quality, low cost flowers from other regions. These export markets are also both further away in distance so that shipping costs are not particularly an advantage. Marketing factors which will affect the attainment of these mid-term goals will likely be related to price and customer service. The capability for the Asian flower industry to initially offer these export market customers low prices, a full production mix, convenient and reliable shipping service will increase the chances of gaining market share in a market place which already has supply.

Lastly, there will be opportunities for marketing cut flowers from southeast Asia to Eastern Europe, particularly northeastern Europe. This would be another likely destination for flowers being shipped out of Kunming. At the moment, economies in northeastern Europe need to improve before significant quantities of cut flowers would be in demand, but by that time (5-10 years) it is likely that transportation routes will have expanded greatly from China to its neighbouring territory, northeastern Europe. As in the past, no doubt winter production of cut flowers in a southern area will surpass cut flower production in the north.

Establishing Domestic Markets

Similar to establishing export markets for cut flowers from southeast Asia, focusing on improving transportation is a key factor to establishing distribution into domestic markets within and among southeast Asian countries. It is proposed that distribution between these countries may be improved if a transportation hub system were enacted. Also, in the case of developing domestic markets, it is likely that efforts to add or improve wholesale markets in major cities would facilitate greater distribution of cut flowers within these countries.

Commercial Floriculture in India

India is a country of small farming households where floriculture is one of the best options for enhancing the income of under privileged. Introduction of floriculture crops could be an important intervention in this regard where a farmer can earn much more by exploiting the available natural resources.

We have favourable climate and cheap labour for growing these crops, whereas they need much less land and water for production. These crops also give premium prices almost round the year and there is no need to wait for a long time as in the case of other routine crops. Net profit against the investment is much higher from these compared with other conventional crops. The products are in high demand all over the world.

Floriculture in the country is in infant stage. There are lack of resources and skilled persons to develop the industry up to international standards. It is the need of the time to produce skilled personals and explore new means to ensure survival of our farmers and explore marketing to save our economy as well as increases export.

The value added products from non-conventional floricultural crops like essential oil of rose, tube rose, jasmine etc., and plants extracts used in medicines and pharmaceutical industry are unique and likely to face less competition in international market after the post WTO scenario and thus have the potential for export and import substitution.

Almost all zones have very rich flora and germplasm which still is not exposed. The available natural resources can be exploited commercially. Marketing of cut flowers is unorganised. In most cities, with large market potential, flowers are brought to wholesales markets, which mostly operate in open yards.

A few flower merchants generally buy most of the produce and distribute them to local retail outlets after significant mark-up. The retail florist shops usually operate on roadsides with little or no protection for flowers. It is the need of time that the government should invest in setting up auction centres, as well as organise floral shops with better storage facilities to prolong vase life of cut flowers.

The production issues which hampers growth of this industry are lack of required seed, flowers, germplasm and tissue culture facilities; lack of technological assistance; shortage of capital investment; shortage of trained manpower; poor post-harvest management; and lack of pest and disease control.

The opportunities existing in this business are proper development of high quality flowers and bulbs for domestic and overseas markets; marketing of essential oils and other products like rose hip syrup, jam, jellies, cosmetics, perfumes and medicines; development opportunities for native flora for domestic and export markets both; identification of market niches that can be supplied taking advantage of the climate; widening of the networking ability of industry and advantage of working with key grower groups to increase market access; and an increase in industry's access to skill-based training schemes.

The challenges which the industry faces are that some popular flower types are over-supplied in international market; lack of quality and quantity assurance of products for maintaining market share in export business; national coordination should be increased to maximise effective promotion in export markets; and the non-availability of skilled labour during harvesting.

Export constraints facing the industry are less encouragement in export of floricultural product with low performance attributed to constraints like the non-availability of air space in major airlines, since most of the airline operators prefer heavy consignments; insufficient number of flights during peak seasons; fear of exporters in regard to infrastructural problems such as bad interior roads, inadequate refrigerated transport and storage facilities; lack of professional backup of delivery and supporting companies, which resort into high cost of technology; and tedious phyto-sanitary certification and an unorganised domestic market.

The government should reduce import duty on planting material and equipment; bring down airfreight to a reasonable level; provide sufficient cargo space in airlines; establish model nurseries for supplying genuine planting material; establish cooperative florist organisations at regional level; set up training centres for diploma course; provide cold chain from farm to destination; and should adhere phyto-sanitary conditions and the WTO laws.

Opportunities for Cut Flower Development

Opportunities for cut flower development in Asia with respect to production are not location dependent, meaning that any area of southeast Asia should be able to capture significant cut flower production opportunity if pursued. In present and future commercial floriculture markets, production opportunities exist if the goal to reliably produce high quality product in consistent quantities can be attained.

To reliably produce high quality cut flowers in consistent quantities requires optimum production management. Production management strategies for implementation and training of personnel will be different for different types of production operations. Varying types of production operations might be commercial cut flower companies, large or small; government directed agricultural farming communities; or rural development programmes managed by public or private funds. For each of these types of production operations, dissemination of pertinent information needed to manage production may require different strategies. Therefore, a key factor to realising production opportunities is education. Ideally, educational resources should be developed and supported by public and private sectors.

An example of educational resources which could be critical to the success of implementing optimum production management for cut flower enterprise in Asia would be a training center/demonstration showcase of production technology adapted to local growing conditions for specific cut flower crops. This educational institution or organisation could facilitate dissemination of information to management at cut flower operations or provide training of technical personnel involved in cut flower operations.

Enlisting the support of floriculture professionals from outside and within Asia to coordinate activities in research and education for the purpose of furthering floriculture development in Asia would have significant impact on the pace and scope of cut flower development in Asia. Formally organised, it would be possible to solicit funds through international agencies to support an institution set up for this purpose, i.e. International Center for Floriculture in Asia.

Departments related to production and marketing management for cut flower crops would be defined in order to reliably produce high quality cut flowers in consistent quantities regardless of the location in southeast Asia. Such departments could be:

— Cultural production, information and management.

— Postharvest production, information and management.

— Equipment use, design, and engineering.

— Genetic improvement of cut flower varieties.

— Consumer marketing.

If sufficient attention and support were given to attaining the goal of reliably producing high quality cut flower product in consistent quantities among Asian countries, then it would be reasonable to expect that the Asian flower industry would soon overtake any other region's flower industry, including the European flower industry, in production. Moreover, given the rapidly increasing rate of spending among Asian consumers for cut flowers, it is also reasonable to expect that the Asian flower industry will soon surpass consumption rates for cut flowers compared to that of other regions.

The European flower industry has led the way in floriculture production, research, floriculture education and training programmes, development of floriculture postharvest handling and storage technology; floral marketing strategies; grades and standards for floral product; and transportation systems for distribution of flowers. Countries within the European region and outside of the region depend on product innovation from Europe to yield new flower varieties. Innovative floriculture production systems have been promoted and exported to other regions of the world. Even floral design concepts have been defined by the European flower industry.

The European flower industry exports more than just flowers. The European flower industry exports flowers, equipment and technology to produce flowers, marketing programmes to increase consumer purchases, and with those marketing programmes, cultural traditions and attitudes about flowers which includes floral design, uses of floral products, i.e. decoration, gifts, holidays, and appreciation for flowers.

Celebration of western holidays and the purchases of flowers for these holidays have become a main theme of floral marketing for consumers in Asia. Two examples of this are cut roses for Valentine's Day and cut carnations for Mother's Day. This type comprehensive involvement with floriculture from disseminating the use of production equipment and technology to influencing consumer purchases of flowers by promoting western holidays associated with floral purchases has established the European flower industry as the leader of commercial floriculture worldwide. Let it be asked, could the establishment of an Asian flower industry spread its influence among countries in Asia and the West? Why not? How would this be done?

There are many opportunities for marketing cut flowers from Asia. In addition to developing a new transportation concept for improving global distribution of cut flowers produced in Asia, there are opportunities in developing marketing strategies which would boost global consumption of flowers produced in Asia. First of all, unifying the image of flowers produced in Asia to create an identity of Asian floriculture would be one step. Then, this could be implemented by introducing an Asian calendar of floral holidays which indicated particular flowers and how they are used in association with these holidays. Even if certain flowers are not presently used in association with a particular holiday or celebration in Asia, there is no reason that floral associations from each country could not invent the concept.

For example, in China, on a certain day in the autumn, the moon and all things associated to be round like the moon are appreciated. Millions of round cakes, labeled "moon cakes" are sold to consumers. Why could not this holiday become a commercial opportunity for the Asian flower industry and pom pom mums become the flower of choice for this day as a gift item along with "moon cakes". If the Japanese celebrate Christmas now with all its commercialism, why could they not celebrate "Moon Day" in the autumn? The same can be said for Americans and Europeans. In Vietnam, flower consumption is auspicious on the 1st and 15th of the month in compliance with Vietnamese traditions. This seems like another convenient way for the Asian flower industry to boost cut flower consumption all over the world twice a month if there were an Asian calendar of floral holidays reminding consumers that it was that time of the month to buy flowers.

Imbedded in this type of floral promotion strategy is the opportunity for the Asian flower industry to develop new varieties of flowers which would characterise these holidays. For example, in case of "Moon Day" not just any type of pom pom mum would qualify, but a particular colour, size, shape, etc. could be developed.

Given the cultural differences among Asian countries, the challenge would be to incorporate as many different cultural traditions representing the different Asian cultures as part of the coordinated Asian flower industry effort. This approach would allow each country to capitalise exclusively on a particular holiday, boosting production and sales for that country according to the holiday celebrated. Once the mission of developing an Asian flower industry was accepted, it would be certain that marketing ingenuity from within Asian flower associations would take over and demand for cut flower products from Asia would soar.

Opportunities in production of cut flowers in Asia would go beyond just the need to increase the rate of production. New production systems for efficiency or environmental protection could be developed which would be adapted to conditions of flower production in Asia. Research priorities at flower research institutions throughout Asia could be directed towards developing innovative solutions which supported global sustainability. Then, in a technology turnaround, it would be countries in the west which would be sending exchange staff to Asia to study new flower production systems. Novel engineering or design applications in equipment and supplies which would be used by producers of flowers for the Asian flower industry could be patented and exported to flower producers in other countries outside of Asia.

With the growth of the flower industry the world over and increased exports, there is need to identify varieties that are most tolerant to the stresses of long shipping and storage. Consumers are seeking products which offer diversity in flower and leaf colour, good texture and form, as well as better performance in interior environments.

Efforts to develop techniques which will reduce the costs of the ultimate product for the consumer will need to be strengthened, with identification of low cost production technologies in protected conditions and refinement and reorientation of imported technology to suit our requirements. Public concern for the environment will require Asian growers to re-examine and change their irrigation, fertilisation and insect control practices. They will also have to reduce the use of pesticides and other chemicals, through effective IPM programmes and increased use of biological control methods. This is important due to the fact that, apart from environmental concerns, developing new pesticides is becoming expensive and governments in several countries are enforcing strict registration requirements for new pesticides, besides removing several pesticides from the market.

Development of the cut flower industry in the Asian region will thus require concerted efforts, not only on the part of the concerned government and growers of a country, but also joint programmes between countries of the region. In view of the commonality of problems and issues, sharing information on initiatives and developing effective networking systems among nations will lead to quicker and less expensive redressals.

Bibliography

Clark D.G., B. Underwood, K. Shibuya, H. Loucas, D. Tieman, and H. Klee, "Microarray Analysis of Floral Senescence in Petunia", *Proceedings of the NATO Advanced Research Workshop on Biology and Biotechnology of the Plant Hormone Ethylene III*, M. Vendrell, H. Klee, J.C. Pech, F. Romojaro (eds.). IOS Press, 2003.

Clark, D., H. Klee and A Dandekar, "Despite benefits, commercialization of transgenic horticultural crops lags", *California Agriculture*, 2004.

Dole, J.M. and H.F. Wilkins, *Floriculture principles and species*, 2 nd Edition. Pearson Education, Inc., Upper Saddle River, N.J. Available in Branson Library, 2005.

Gentile, A.G. and D.T. Scanlon, "A guide to insects and related pests of floriculture crops in New England", University of Massachusetts Cooperative Extension System, 1992.

Klee H.J. and D.G. Clark, "Ethylene signal transduction in fruits and flowers", *In: Plant Hormones: Physiology, Biochemistry and Molecular Biology*, (3rd ed.)", PJ Davies (ed.), Kluwer Academic Publishers. Dordrecht; London, 2004.

Larson, R.A., *Introduction to floriculture*, Academic Press, N.Y. [Branson Library], 1992.

Negre F., C.M. Kish, J. Boatright, B. Underwood, K. Shibuya, C. Wagner, D. Clark and N. Dudareva, "Regulation of floral scent emission after pollination in snapdragon and petunia flowers", *The Plant Cell*, 2003.

Powell, D. A., and T.S. Bellows Jr., "Development and reproduction of two populations of Eretmocerus species

(Hymenoptera: Aphelinidae) on Bemisia tabaci (Homoptera: Aleyrodidae)", *Environmental Entomology*, 1992.

Russell, Maxine K., *Leaves from a Greenhouse*, 1984.

Schuurink R.C., M.A. Haring and D.G. Clark, "Regulation of volatile benzenoid biosynthesis in petunia flowers", *Trends in Plant Science*, 2006.

Seeley, J.G., *Advancements in Commercial Floriculture*, Southern Florist and Nurseryman, 1982.

Szabo, P., J.C. van Lenteren, and P.W.T. Huisman, "Development time, survival, and fecundity of Encarsia formosa on Bemisia tabaci and Trialeurodes vaporariorum", *SROP/IOBC WPRS Bulletin Working Group IPM Glasshouses*, 1993.

van de Veire, M., "Side-effects of pesticides on the parasitic wasp Encarsia formosa: comparison of results from laboratory testing methods with practical experiences in glasshouse vegetables", *SROP/IOBC WPRS Bulletin*, Pesticides and Beneficial Organisms, 1994.

van Lenteren, J.C., and J. Woets, "Biological and integrated control in greenhouses", *Annual Review of Entomology*, 1988.

Index